Jason Wilson, Emerit... London, has written books on C... Neruda, Alexander von Humboldt and travel companions on Latin America, the Andes and Buenos Aires. He spends his time between London and Buenos Aires.

Also by Jason Wilson

Octavio Paz: A Study of his Poetics

Octavio Paz

Traveller's Literary Companion to South and Central America

Buenos Aires, A Cultural and Literary History

A Companion to Pablo Neruda

Jorge Luis Borges

The Andes, a Cultural History

Buenos Aires

Edited and Translated

Alexander von Humboldt, *Personal Narrative of a Journey to the Equinoctial Regions of the New Continent*

Octavio Paz, *Itinerary*

Jacques-Henri Bernardin de Saint-Pierre, *A Journey to Mauritius*

Jorge Luis Borges and Osvaldo Ferrari, *Conversations*

LIVING IN THE SOUND OF THE WIND

JASON WILSON

CONSTABLE • LONDON

CONSTABLE

First published in Great Britain in 2015 by Constable
This edition published in 2016 by Constable

A CIP catalogue record for this book
is available from the British Library.

ISBN: 978-1-47212-205-6 (paperback)
ISBN: 978-1-47210-634-6 (ebook)

Typeset by SX Composing DTP, Rayleigh, Essex
Printed and bound in Great Britain by CPI Group (UK) Ltd, Croydon CR0 4YY

Papers used by Constable are from well-managed forests and
other responsible sources

MIX
Paper from
responsible sources
FSC® C104740

Constable
An imprint of
Little, Brown Book Group
Carmelite House
50 Victoria Embankment
London EC4Y 0DZ

An Hachette UK Company
www.hachette.co.uk

www.littlebrown.co.uk

For Andrea

Contents

'One does not want to recommend it as a book so much as to greet it as a person, and not the clipped and imperfect person of ordinary autobiography, but the whole and complete person whom we meet rarely enough in life or in literature.'

Virginia Woolf reviewing W. H. Hudson's
Far Away and Long Ago, 1918

'To know that you could read me is good news indeed – for one writes only half the book; the other half is with the reader'.

Joseph Conrad in a letter to Robert Cunninghame Graham,
5 August 1897

Chapter 1

Traces of Hudson (1841–1922)

The work and persona of W. H. Hudson – at one time one of England's best-known novelists and naturalists – have drifted into a timeless byway and to think of him is to enter a pious, sentimental zone. He is held in such awe that his work is almost beyond literary criticism. You love him or ignore him. Most readers ignore him, unless the mere mention of his name elicits a fondness that transcends literary categories. But if you are an Argentine, you might consider him one of yours, translate his initials W. H. into full Christian names Guillermo Enrique, and read him in Spanish. Even then, his main work, *Allá lejos y hace tiempo* (*Far Away and Long Ago*, 1917), with as catchy a title in Spanish as in English, is one of those books given as school prizes that remains unread. I had often been tempted to update Hudson, to rescue him from his admirers and make him modern, but I hadn't gone beyond accumulating his fragmentary works, casually bought from second-hand bookshops.

His conflictive identity as an Argentine, born into a displaced American family, and as an Englishman, had appealed to my own dilemmas. I have been puzzled by place and where I belong. As a Mauritian born from a Mauritian family, with a Norwegian mother and settling in London as a seven-year-old, I have struggled all my life with where I really belong. And the problem gets knottier. My Mauritian-born father Frank Avray Wilson turned his back on the tropical island of his own birth – in fact he hated it – and launched himself into the life of a *tachiste* or abstract expressionist painter in the London of the 1950s. He had trained at Cambridge as a natural scientist and tried, intellectually, to heal the breach between art and science (as he liked phrasing it in such general terms) by writing several books about this topic. Hudson too turned his back on Argentina, and lived and wrote between the scientific and artistic worlds. Nobody in my family ever saw themselves as immigrants or exiles. Coming from a colonial society like Mauritius had promised an imaginary England as spiritual home. Hadn't Hudson carried out the same promise with his coming to England? Nevertheless, I only saw this lure of a dream England clearly as I explored Hudson's roots and his adopted Englishness. For a long time I thought, like my father, that I was English.

It wasn't odd that my parents really preferred living in France, where they moved in the 1960s. Mauritius was as French as it was English. But it was mother's life that added to the identity conundrum. After her death in 2008 I discovered that she was illegitimate and born in Stockholm, not Oslo, as all her documents had stated. She had hushed all references to her birth and deliberately ignored her birth mother. All I knew about her was her date of birth in the parish registry in Stockholm, which I managed to locate two years after her death. There's much family speculation about who our grandmother and the anonymous grandfather were. Yet our mother could have met her birth mother, but declined. Many families have gone through these suppressed dramas and there's nothing unique under the sun, only her decision affected her four children. Instead, she became attached to her English, aristocratic stepmother Alice Higford and her class values. So that's why 'England' became the home my parents dreamed of until they actually arrived in London in 1950.

In fact the cautious assimilation of my parents into England and its class values was completely silenced. As was Hudson's in the late 1870s. So, through my work on Hudson, I began to understand these dumb-dramas of belonging, with no idea that they would continue into the future because they'd been denied. I even contributed by marrying Andrea, who came from Argentina. We had three daughters and thus pushed my identity puzzle even further into the metaphorical dark pit. I sensed that behind Hudson's casual prose there was a silenced sub-world and that his root confusions somehow mirrored my own.

My approach to the puzzle of identity, I see now, has been very bookish. I'd become a university lecturer, had written books on Octavio Paz, Jorge Luis Borges and Pablo Neruda and absorbed the surrealist adventure. But, like all closet surrealists, I was aware that the world did not end up as a book. My impatience with scholarly learning ensured I never completed my doctorate. It could be as simple as my growing up with the US Beats, the French and Latin American surrealists and Camus and Sartre. I was a proud cosmopolitan. England in the early 1960s seemed parochial and absorbed with tradition, class and national history and I sought a new-world freedom. So even if pampas-born Hudson is an alien, far-off figure, locked in the remote past, I sense he holds some clues to our twenty-first-century dilemmas. After settling down with my Argentinian wife, I set about exploring his confusions in both the land of his birth and his adopted country.

I was accompanied by the usual contradiction: he was a meticulous fiction and natural history writer and yet, as a man, towered above his works. Morley Roberts reckoned that he was first a man, then a writer; that Hudson, his friend, was immeasurably greater than his books. Of course, he would say that. But maybe he's right. A biographical approach to this self-effacing naturalist suited me down to the ground. I would track down this elusive writer.

Before turning to books and archives, I wondered what traces W. H. Hudson had left physically behind him in England. I began in Buriton after discovering a W. H. Hudson walk on the web and downloading it. In 1900 he'd found this 'small pretty rustic village in a deep hollow among the downs where Gibbon was born'. I was curious how the area had changed since then. I don't have a car, so boarded a train at Clapham Junction that would take me to Petersfield. In Hudson's day, steam engines made Clapham Junction dirty and busy, so that he could use it as a metaphor for the worst of urban life. It's still hectic. But Petersfield awaited as another England. Its large main square, with Norman church and equestrian statue of William III, teemed with white English people.

A local, empty bus took me to Buriton, stopping on the way at a massive Tesco and then at a bungalow village for the old. Dropped off, I strolled down to a duck pond with alders and lime trees, another small Norman church and a large, fortified farm, three elements that withstood change. Edward Gibbon had been born there, though no plaque stated this, and Gibbon's six-volume *Decline and Fall of the Roman Empire* had been part of the Hudson's pampas library. The dreamy duck pond's surface broke with feeding carp and three sullen, unemployed youths were trying to fish them. A family fed white sliced bread to the ducks.

I cut up a steep wheat field into a wooded hangar. Sunlight fell, dappled, through the leaves; there were blue tits and cuckoo spit

flowers. Two colourful walkers in proper gear walked earnestly past. Later, after my walk, I met them at the bus stop waiting for the last bus at 4 p.m. We chatted. They were also finishing a 'literary' walk, following Edward Thomas's trail. I mentioned my Hudson research and his friendship with the poet. The man, a retired doctor who looked like an actor, nodded: he too had downloaded the Hudson walk. It dawned on me that Hudson would have despised us. He would have loathed 'literary' walks, in fact any kind of guided tour. He would have wanted the wilderness of the Buriton hangars to remain wild. The old, overgrown quarries with plaques explaining their history would have depressed him.

Over the period he roamed southern England, from the 1890s to the 1920s, this labelling of the countryside hadn't existed. England hadn't been turned into a National Trust park. He even hated Kew Gardens because every tree was tagged. He sought areas of 'isolation and loneliness and unchangeableness'. He would have despised not just 'literary walkers', but even hikers and people with dogs.

Luckily, not a sign indicated a Hudson walk. He would have strayed into dense wood, escaped us ramblers as if we were lepers. I was reminded of Gerard Manley Hopkins's lovely last line to the poem 'Inversnaid': 'long live the weeds and the wilderness yet'.

My next venture in tracking down Hudson and his desire for 'isolation and unchangeableness' (his awkward word) was the headquarters of the Bird Protection Society he helped found, which became a Royal Society in 1904. The RSPB moved to its current premises at Sandy, Bedfordshire, in 1961, so Hudson could not have known the large Victorian mansion, employing some 600 people in a grand park being developed into one of native trees for native birds.

I spent the whole day examining boxes labelled 'Hudson' in their library. The RSPB had inherited his papers, or what he hadn't managed to destroy or burn. My feeling was one of déjà vu. Earlier biographers like Ruth Tomalin and Alicia Jurado had combed the

dusty papers thirty years before. Meanwhile, a friend, who would join me for several trips in my search for Hudson's traces, wandered the park and its guided tracks in light drizzle, crossing quiet bird-watchers with binoculars. There was a spirit of peace and gentleness. Then the young, cheerful librarian wanted to show us the Board Room, so we followed her into a large room, and above the fireplace there it was, an oil portrait of Hudson by Frank Brooks. Writer and artist never met, so it was painted from a well-known photo of Hudson in the New Forest. It had cost the society £200.

I'd felt, up to that moment, that Hudson had been relegated to unread boxes, but here he was at the heart of the building, in a section closed to the public. His guiding spirit was alive. The society's purpose to protect birds thrives. It's Europe's largest wildlife conservation charity. It's also where Hudson fell passionately in love.

To keep Hudson's spirit alive is what his commemorative plaque promises, if you can read it. Hudson lived in different lodgings in

W H HUDSON
WRITER & NATURALIST
1841 - 1922
In commemoration of the 150th
anniversary of his birth
1991
Mario Campora   Charles A. Muller
AMBASSADOR   CHAIRMAN
ARGENTINE EMBASSY   ANGLO-ARGENTINE SOCIETY

the North Kensington and Paddington area of London from his arrival in England in May 1874 to his death in 1922. On a corner of 40 St Luke's Road, his London home from 1886 to his death, Hudson's Friends Society of Buenos Aires had placed a square plaque in 1938. It's not blue, but a bronze relief by the well-known Argentine sculptor Luis Perlotti of Hudson's birth shack, called Los Veinte-cinco Ombúes (Hudson's old-fashioned spelling), near Quilmes. It features an engraved line from near the start of *Far Away and Long Ago* that reads, 'The house where I was born in the South American pampas . . .' He's also referred to as a 'great writer', but you can hardly decipher it.

On the same house, a second, more readable plaque was placed in 1991 by the Argentine ambassador and the Anglo-Argentine Society, but now he's lost the 'great' and become just a 'writer and naturalist'. Again, it's not blue. Each time I'm there I promise Hudson I'll clean up the plaques. The Argentines are keener in claiming him as their own.

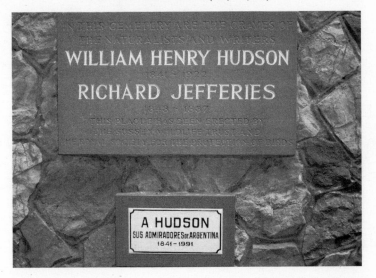

Another monument is Hudson's tomb in Worthing's Broadwater Cemetery. He is buried with his English wife, Emily. From the station we walked up a quiet street, lined with mock Tudor red-bricked houses, until we reached the cemetery. Before going under the gatehouse, there's a little fenced-in garden, with yews, wildflowers and birdbath, dedicated to Hudson and to fellow naturalist and novelist Richard Jefferies. Signs directed us to the simple tomb, where the cheap cross has been left fallen. Hudson's friend Robert Cunninghame Graham had tried to prevent the erection of any cross. The tomb lay unkempt by the road and a holly hedge. It was in stark contrast to Jefferies's tombstone, kept in good order by the Richard Jefferies Society. Hudson had asked for simple daisies to be planted there, but there aren't any.

You can read his full name, William Henry Hudson, his complete birth date, 4 August 1841 in Buenos Aires, and his death date, 18 August 1922 in London. Below, Emily his wife is given just her death date, 19 March 1921. He ended up buried in Worthing because his

In Memory of
WILLIAM HENRY HUDSON
BORN 4TH AUGUST 1841 IN BUENOS AIRES
DIED 18TH AUGUST 1922 IN LONDON

Also in Memory of EMILY, his Wife
DIED 19TH MARCH 1921 IN WORTHING

invalided wife had died there one year before and also to be close to Richard Jefferies. Nobody had cleaned his tomb; it was far from clumps of lovely pines and had no birdbath. It was the oblivion Hudson wanted.

Someone had decided on the lines 'He loved birds and green places and the wind on the heath and saw the brightness of the skirts of God', fusing a line from George Borrow with some from the little-remembered US poet William Cullen Bryant (1794–1878). The reference to God is wrong. Hudson didn't believe in the afterlife or in personal immortality; he knew from a boy that nobody survived death.

Standing at his neglected tomb and reading the epitaphs, I realized only seventeen months separated the deaths of husband and wife. He'd pondered how old couples followed each other to death, with the survivor's 'awful solitude' a spur to die. It happened to my old parents in their early nineties. Grief for the death of his wife left my father without reason, lost in dementia. He survived her by nine months of hell. We were the only people in the cemetery. You can

spot the downs and the Chanctonbury Ring Hudson loved. His spirit was up there, though he had also sauntered the cemetery itself as an old man. When Henry Williamson visited Jefferies's tomb in early 1921, a gardener told him 'that an old gentleman with a beard and the eyes of an eagle told him he would like to be buried near Jefferies when his time came'.

The most noteworthy monument recalling Hudson is the Hyde Park Bird Sanctuary in London. Few park strollers realize it's dedicated to Hudson, or who he was. In fact, it's marked on the park maps, but not in the latest *A–Z London* maps. I've taken many people there to stare in surprise at Jacob Epstein's solid Portland stone block with its relief of an erotic, primitive Rima flying up, hair streaming, bare-breasted, among mythic birds. Epstein's first maquette was of a gangly Hudson lying back in a grove and bird-spotting, but the park authorities refused to have a representation of an actual person. So Epstein picked on Rima, his humming-bird character from Hudson's romance, *Green Mansions*. Eric Gill was also involved and had incised Hudson's dates in roman numerals on each side of the relief.

Epstein's *Rima* took seven months to carve in his studio in Epping Forest. Roger Fry, Bloomsbury's art critic, found her 'with the haggard shyness and strangeness of wild things'. Swiss-born Aimé Félix Tschiffely (1895–1954), biographer of Hudson's friend Robert Cunninghame Graham, claimed Hudson used to kip in that exact spot in Hyde Park when lonely and poor. Tschiffely knew a little about Hudson's wild spirit: he who rode two horses, 'Mancha' and 'Gato', from Buenos Aires to Washington over three years.

In a corner of the sanctuary, a black, streaky slate tells us that it is dedicated to the memory of Hudson, 'Writer & Field Naturalist'. Then along the border of the stone bird troughs there are more chiselled references to W. H. Hudson. It's railed in, but is a quiet spot, with a bench where you can sit and reflect. As Tom Sutherland showed me, it also once had a fountain that worked.

Nowhere does it reveal the scandal of its unveiling on 19 May 1925. It's funny how history revitalizes a monument. Cunninghame Graham had argued for avant-garde Epstein to design a bird sanctuary in Hyde Park (after all, Epstein had sculpted his own bust just before). When Epstein was nominated, according to Linda Gardiner, who was on the committee, Lord Grey and John Galsworthy, two Hudson defenders, resigned in protest. First move of the anti-Epstein cohort.

Finished, the Hudson monument was unveiled by Stanley Baldwin, the Prime Minister, who later collected his speech in his essays *On England* (1926). I've seen a photo of Baldwin standing before the unveiled block, not knowing what was to appear. Hudson, he said, came from the ends of the earth and made the hidden beauties of the southern counties familiar to thousands. He listed Hudson's bêtes noires – his disgust at the felling of the elms with its

Jacob Epstein by his *Rima*

rookery in Kensington Gardens, eating plovers' eggs and bird feathers decorating women's hats.

Then, after a speech from Cunninghame Graham, the promoter of Epstein, Baldwin unveiled the bird sanctuary. There was a gasp of horror from the crowd. Naked Rima became headlines in the newspapers and generated furious letters to *The Times*. A 'hullabaloo' is Philip Gosse's word for it. In a letter to the *Morning Post*, signed by Hilaire Belloc and Sir Arthur Conan Doyle, amongst others, *Rima* was deemed 'inappropriate and even repellent'; the writers wanted it removed. Newspapers nicknamed it the Hyde Park Atrocity. Hudson's friend, Margaret, Lady Brooke, the Ranee of Sarawak, agreed and wrote, 'Get the FOUL thing removed'. The scandal even reached the House of Commons where Sir Basil Peto requested the removal of that 'specimen of Bolshevik art'. It was daubed and desecrated, and in 1935 a swastika was painted over it in what Epstein called national Rimaphobia. It made Roger Fry issue a call to all who cared about art to 'face the modern foe', the delicate Philistines, even if he didn't like Epstein's sculpture.

Another trace of Hudson is a simple granite stone at Zennor, Cornwall, with 'W. H. Hudson often came here' carved on it, still just decipherable. Zennor, a lonely village nestling among the furze, was near where Hudson wintered from November 1905 onwards. He would sit on Zennor Hill, by the old quarry, and survey the desolate scenery. According to a cutting from the *Cornishman*, a Mr Lewis Hyde often saw Hudson sitting on this boulder in what was known in the 1920s as Hudson's Seat. Whenever Hyde tried to approach him, Hudson refused to answer, so absorbed was he in his thoughts.

Unable to sleep one night at Zennor, Hudson had an insight about 'this dreadful unintelligible and unintelligent power that made us', what you could call the terrible aspect of nature. A divine indifference characterized this 'all-powerful and ever-lasting, creator and slayer of all things that live, of beauty', that was 'without knowledge or thought or emotion and that which he made and would unmake was without significance to him'. But, though melancholic, he was not pessimistic and hope came with dawn and starlings.

At the church in Martin, some eight and a bit miles south of Salisbury, there's a little plaque mentioning Hudson. We hired a car and drove there. Up a narrow lane we found the humble village church that Hudson loved. He would always visit the churches in remote villages during his rambles, and churches hardly change. After parking the car under an enormous apple tree with green cookers, we immediately saw the tombstone of William Lawes, shepherd father to the shepherd who featured in Hudson's *A Shepherd's Life*, in front of the church. It was almost as if the tombstone had been moved there so you wouldn't miss it.

Three different villagers we talked to, like the Lawes family, knew that Hudson had written about Martin, thus perpetuating this shepherd. Hudson was known as an author who'd immortalized their village in a book. But Martin was desperately poor at the turn of the nineteenth century. Now it's pretty, with climbing roses and well-kept lawns, an estate agent's dream. Sitting on a bench in the cemetery, eating our sandwiches, we could still glimpse the downs and the isolation Hudson so craved.

A last English memorial to Hudson is Stone Hall Sanctuary on the Easton Lodge Estate of Frances, Countess of Warwick. I had to visit it and see if the lodge still existed, as no biographers had mentioned it. Before taking the train to Stansted airport and then a taxi in squalls of rain, I found out that Daisy Greville, society beauty married to Lord Warwick, had led a tempestuous life, even being one of Edward VII's mistresses. But, unpredictably, during the war, this gossipy nature lover veered, scandalously, into becoming a sympathizer of

William Lawes tombstone, with small plaque at
bottom mentioning Hudson, dated 1919.

the communist October Revolution and then the Labour Party. She
tried to donate her enormous house to the Trade Union Conference
in 1923, but then it burnt down, losing twenty-eight bedrooms.
According to a biographer, one of her pet monkeys dragged a blanket
across a fire and set the blaze going, burning all her papers.

As an amateur, she wrote on nature and on the socialist William
Morris, and was a close friend of H. G. Wells, who lived at Grebe, a
house nearby. Her 800-acre park became a sanctuary for her horses
and all wild life, and she banned fox hunting. She invited the
aristocratic eccentric Robert Cunninghame Graham to open the
Hudson house, Stone Hall Sanctuary, in 1923 in the estate's Garden
of Friendship, with a library of Hudson's works. H. J. Massingham,
another nature writer and Hudson fan, also attended. 'I hope and

believe,' she wrote, 'that many students will spend happy hours in the sylvan seclusion'. But she was too capricious to upkeep the memorial, and after her death it vanished. During the war, the United States Air Force cut down 10,000 oaks to make an airfield. The house fell into disuse and was demolished. The zoo was shut down and the monkeys were shot. Countess Warwick claimed Hudson as 'one of the rare intellects of my generation'.

When we reached the garden's entrance outside Great Easton, a sign said it was closed to the public. I took the taxi's number, sent it back to Stansted and we climbed over the fence into the garden. Little remained of that grand lady's whims but some upright columns, a gazebo, flowers beds and evidence of landscaping among the trees in the rain. Walking round what was the sole remaining building, a wing of the great pile, we bumped into the current owner walking his dog. We had been caught trespassing, but he liked my story, though had never heard of Hudson. We gossiped about Lady Warwick and he walked us to Great Easton village duck pond and church from where we could, with luck, get a bus back. One detail I retained about Countess Warwick was that she insisted on being addressed as Comrade Warwick. All obeyed except her servants who continued calling her 'your ladyship'.

So much for Hudson's meagre physical traces in England, but I also want to check what's happened to Hudson, that is Guillermo Enrique Hudson, in Argentina. It's almost as if he's another person out there. I came across his work as I was getting to know my wife's Argentinian family in the 1970s. We both came from landed families on the wane. Nothing could be more dissimilar from a small Mauritian sugar estate than a cattle farm on the pampas, but our sensibilities met at that point of the arc. Poet Alberto Girri had warned me not to translate already known poets. 'Be the first to translate,' he said. I was on the look out to find a guide to my new life and found Hudson. We were on reverse journeys. He had gone to England and I was going to Argentina. Few had written on him and I loved his calm writing. There were further points of contact, not least Hudson's own passage between science and art which mirrored my own often antagonistic arguments with my father. As I became familiar with farm life, from castrating bullocks to having *mate* with farm hands, I became familiar with a part of Hudson's life that was out of most people's range. Life on my wife's cattle farm was essentially nineteenth-century in comparison with the modern island-city of Buenos Aires. In fact I was privileged to see this farm activity in its last years before television, combine harvesters and mobile phones took over. Hudson was also an author in my mother-in-law's small library as a field naturalist. She was a keen member of the Argentine bird society and had read him in Spanish. I slowly built up a patchwork of experiences of the pampas that I could test with Hudson. I even found a reference to my wife's land in Hudson's writings. The big farms have been broken up and divided by family successions, but Hudson had fixed its peculiar life-style in time and in print.

Drivers speeding along the motorway from Buenos Aires to La Plata, the provincial capital, stop and pay at a *peaje* (tollbooth) called 'G. E. Hudson'. Nearby, there's a train station called Guillermo Enrique Hudson (before 1930 it had been called Conchitas, the name of a stream that passed through the Hudson land). I bet that 99.9 per cent of its users haven't a clue to the identity of the name.

Despite a map, the most recent time we drove to the Museo Provincial Guillermo Enrique Hudson in Florencio Varela, Quilmes, a regional borough of Buenos Aires, we got quite lost. Bumping down mud streets with precarious housing, but not shanties, we drove in circles. I had to ring Rubén Ravera, the director of the Hudson museum, to see if he'd come and fetch us. We met at a petrol station and he guided us to our destination. Once, he told us, he invited the English and Japanese ambassadors there, to thank them for donating money to buy more land, but they couldn't get there. When it rains the dust roads turn into impassable mud. By then I had met Rubén several times, and he's wonderful value, and a canny defender of Hudson's writings.

The museum's first director, Violeta Shinya (1910–2003), was Hudson's grand-niece. Her mother Laura was the daughter of Hudson's sister, Mary Ellen, and married the Japanese immigrant Yoshio Shinya. Violeta, after her mother's early death, was brought up by Mary Ellen and inherited Japanese looks from her father. I met her once. She sat in an armchair with a rug and chatted about the difficulties of translating Hudson into Argentine Spanish. She died without descent. She was the closest I came to Hudson in person, though there is one descendant with the Hudson surname.

Rubén Ravera had taken over from Violeta Shinya when she retired in 1990 and then fallen out with her. She was an elitist who lived under the shadow of the well-connected Argentinian writer Victoria Ocampo; she was also jealous of Hudson's biographer, Alicia Jurado. Rubén Ravera's dream is to write a history of the Museo and his relationship with Violeta Shinya. But he's very *porteño* (as Buenos Aireans are called), and he's a master of many trades promoting Gesell's economic philosophy, admiring English-born E. F. Schumacher and studying to become a museologist. He once had a job bringing ice to an Amazonian tribe on the Río Negro in Venezuela, and another organizing a barter market in Caracas, and that's just a start of the many balls he's juggling. But above all, he loves Hudson and does all he can to stock a library and preserve Hudson's stature, although he once bitterly said to me, 'I've nobody to talk with about Hudson' (biographers Alicia Jurado, Ruth Tomalin and Dennis Shrubsall had all recently died).

Los Veinte-Cinco Ombúes today.

Inside the grounds of the Museo, there's a bust of Hudson, a poor likeness, and his shack. I am always moved when faced with what's left of his home, testimony to the Hudsons' poverty. W. H. Hudson's Los Veinte-Cinco Ombúes, where he was born on 4 August 1841, was a humble '*rancho*' or small farm house, dating from 1750, the oldest building in the region. He was born in this peasant's

Hudson's birth shack in the 1930s.

thatched shack, with just three small rooms, a protruding roof functioning as a gallery and no bathroom. It was made with thin, sun-baked adobe bricks, with an adobe-brick floor. There are no stones in the pampas.

How this large family lived in such cramped spaces is a puzzle. Did the children sleep together, with the parents in a second room? Did the sole girl (a second sister was born later in Las Acacias) sleep with the parents? According to Rubén Ravera, the kitchen was a separate house, as was the washroom, but neither stands today. The first time I visited, I had wanted to know if the kitchen was any different to the typical gaucho one described in Hudson's tale 'Niño Diablo', lighted by three wicks in cups of melted fat and heated by a great fire in the middle of a clay floor. All around, on hooks, riding gear and kitchen utensils. At the fire, beef would be roasting on a spit and a large pot would hang from a beam filled with mutton broth, with a smaller pot for extras. Everybody sat around on chairs or stools. Kitchens were the hub of social life, so the building's absence denied me a glimpse of the young Hudson's life. I'm left without a precise idea of how the Hudsons lived. I'd visited a farm-worker's mud-floored hut in the 1970s and knew of several nineteenth-century prints like Pallière's, but none could give me an exact image. The Hudsons were North Americans in exile and just very different and more cultured, yet they were as poor as their neighbours.

Despite the squashed space for a family of eight, there was a library of some 200 volumes. This pampas library was the decisive factor that distinguished Hudson from the neighbouring, illiterate gauchos. The books have been dispersed, but Ravera and his team have amassed an excellent, new collection, housed in a special hut. A year earlier, I had spent a day going through all their Hudson documents. There's also a run-down café, where you can buy

Jean Leon Pallière, 1864.

Hudson in translation. In the place of honour, there's a photo of late president Néstor Kirchner, holding a copy of Hudson's *Allá lejos y hace tiempo* in his hand. The little ecological reserve survives by hosting events and talks.

It took Dr Pozzo, mayor of Quilmes and a doctor, two years to locate this *rancho*; after much searching of landholding documents, he found it in 1929. He decided to translate *Far Away and Long Ago* into *Allá lejos y hace tiempo* in 1938. Just before his death in 1950, he persuaded the Davidsons to donate the four hectares on which the shack stood. This birth *rancho* became the Guillermo Enrique Hudson Museum in 1957. In 1980 it changed its name to the Parque Hudson para la Ecología y la Cultura, run by the Sociedad de Amigos de Hudson.

The first time I reached the Museo in 1987, driving round and round the poor area, I discovered that what was a Parque Ecológico on my map was actually a football pitch, covered with tattered plastic bags. Finally, a padlocked gate led onto the Hudson family land through a wood of black acacia and paraíso trees. A man on a horse

Jean Leon Pallière, *El Ombú*, 1864.

suddenly loomed up and opened the doors to the hut. Inside the small rooms were books, photocopies and much about Cunninghame Graham. It was like a messy schoolroom.

The name of the shack, Los Veinte-cinco Ombúes, came from an avenue of twenty-five ombú trees (although the ombú is hollow inside, more like a giant thistle). The evergreen tree had great significance to Hudson. The sole surviving ombú on the site could be 'over a hundred years old', said the laconic horseman. The shack, with a corrugated tin roof, was built on a rise, a loma, and looked on to the Las Conchitas stream and fields of enormous drying thistle, which was the main fuel for the kitchen. Hudson remembered how in certain years, the spiky thistle grew so high you couldn't ride out of your house; he called it a thistle year.

Years later, current director Rubén Ravera drove me there from Buenos Aires for my second visit. Before setting off, he pointed to a house a few doors down from where he lived in the south of Buenos Aires on Bolívar street and said that Hudson used to stay there on Bolívar, 314. It belonged to the Methodist minister, and Hudson recalled playing there in a covered patio. The original colonial house was single-storeyed, with inner patios, at the very edge of town, with a stream along the bisecting street. How amusing, I thought, to take the same thirty-three kilometre journey in a van that Hudson had taken on horse.

At the time, Ravera was expanding the museum's land into an ecological park. With donations from Lloyds Bank and Japanese funds organized by Matsuo Tsuda, the ex-ambassador to Japan who ran a Hudson foundation in his homeland, Ravera had bought up 54 hectares. He was constantly fighting land grabbers, marauding cows and hunters who saw free wasteland and not an ecological park. Squatters had built a corrugated tin shack and refused to budge. When two permanent policemen were lodged there, tensions calmed.

The huge, single ombú I had seen on my first visit still stood as a sentinel to Hudson's days. An alert naturalist-guide confirmed that it was hollow. He took me on a tour of the park with a tripod and special binoculars, and had counted forty-two different bird species that morning. The direction in which the ebullient Rubén was leading the Museo was exactly right, an ecological centre, near enough to Buenos Aires to function as a kind of closed-off sanctuary or paradise for urbanites. You didn't have to read Hudson to appreciate this spot, surrounded by factories, gated communities and shopping malls, some abusing the name Hudson, and the inevitable shanties. It is a living monument to the kind of ecology that Hudson revered.

Before my first meeting with Rubén Ravera I decided to take a trip to the town of Quilmes itself, now a suburb to Buenos Aires, to see a statue of Hudson by sculptor Santiago Parodi. Quilmes was named after the Kilmes who lived in a Valle Calchaquí fortress city, near Tucumán in the north of the country. They resisted the Spaniards until they were defeated in 1666. They were relocated

The Ombú at Hudson's musem.

fifteen kilometres outside Buenos Aires. Quilmes is now famous
for its beer – still called Quilmes – and an English public school,
St George's.

I took a 98 bus called a *colectivo* or *bondi* in Spanish from a plaza.
Once taken over by street sellers and slow-walking indigenous
South Americans. It covered the thirty-odd kilometres in an hour
and a half of constant traffic lights and identical, low-level, concrete
and shabby buildings, with neon signs and second-hand car
salerooms. I could be anywhere in what used to be called the Third
World. I didn't know when to get off, it was so numbingly
monotonous, but a woman sitting next to me pointed out the
station plaza, and I jumped up stiffly.

I walked seven blocks to the neo-colonial cathedral with its twin
towers. Hudson's mother had contributed to its building in 1858 (it
was finished in 1866), just as Hudson himself had given something
towards paving the road from Quilmes to his shack. Then I found
the Monumento a Hudson, standing amongst trees on the Plaza del

Bicentenario. His name was engraved in the pedestal and he was described as a 'naturalist and writer of Quilmes' in Spanish. Santiago Parodi, after travelling Latin America as a professional boxer, reinvented himself as a self-taught sculptor. He spent two years on Hudson's bust.

In 1940, the discoverer of Hudson's hut, Dr Pozzo, organized the unveiling of Parodi's statue. It was the first public Hudson event in Argentina and celebrated the century since his birth. Pozzo said: 'After a hundred years, Hudson has returned to being an Argentine.' The bust stood tall, gilded in bronze. It was a good likeness. Months after my visit, I discovered that Antonio Parodi, Santiago's brother, had painted twenty-six oils of the Hudson shack. Parodi is my wife's surname and at an auction we had bought two Parodi landscapes, without knowing who on earth he was.

After looking at the statue, I wandered into the main Quilmes bookshop and asked for any books by Guillermo Enrique. I was told to go to a smaller bookshop on the other side of the cathedral square. I'd see Hudson's books in the window of a tiny second-hand bookshop. I identified myself to the owner, who turned out to be the sister of Rubén Ravera, whom I hadn't yet met. This happy consequence of an almost pointless journey led to my meeting him and being driven out to the Museo Hudson.

Another day-trip needed to be organized. When Miriam, a yoga-teacher friend, offered to drive us, my wife and I leapt at the chance. There were two versions of where the Hudsons had lived in their house, Las Acacias, from about 1846 to 1856, after they had moved from the shack. Hudson remembered a vague, day-long cart journey in June, winter, to the new, rented home Las Acacias (named after a tree) on the route to Chascomús, where most of *Far Away and Long Ago* happened. The Hudson family moved by large-wheeled cart, drawn by oxen, along the Camino Real southwards in the direction of the village of Chascomús. Hudson's memory may have shrunk this journey, as an early Hudson biographer, Jorge Casares, had asked an old man who had travelled the same distance in 1874 and was told he set out at three in the morning and arrived late at night.

Las Acacias was a thatched adobe-brick house with high ceilings and large windows with iron bars to let cool air pass through. Behind this large house that fronted the road was a store, a dairy, a kitchen, a dovecote, huge barns and woodpiles of thistle stalks. There was also a seven-acre wood of pear trees, black acacias (with sharp thorns) and Lombardy poplars, but no pines, no eucalyptus, no evergreens. This was Hudson's childhood 'wooded wonderland'. The site was originally a small frontier fort. A moat, some twelve feet deep and thirty wide to defend from Indian attacks, surrounded the buildings.

But where was it? Jorge Casares was sure that it was the old Estancia Vitel, a ranch or farm belonging to Leonardo de la Gándara. The Gándaras had been exiled by tyrant Juan Manuel de Rosas to Montevideo, and so would have taken any tenant. They returned in

1856 after Rosas's defeat in the battle of Caseros. Gándara was a judge in nearby Chascomús, and a Gándara appears in *Far Away and Long Ago*. Remains of the irrigation ditch of this Estancia Vitel were found, Casares claimed, near the railway station of Gándara. The station was established in 1865, long after the Hudsons had left. We drove there and walked up and down the platform and around the lovely station. The Mar del Plata express passed through. But not a sign of the Vitel house or of any ditch.

We were having a picnic on the front lawn of an abandoned worker's cottage, opposite a large, fenced off and closed-down factory, when a guardian approached. After the inevitable formalities, he suggested I call on a sixty-five-year-old who might remember, but I declined politely, aware that personal memory was no good. Hudson had lived there over 170 years before. The factory hadn't existed then. It made the well-known Gándara *dulce de leche*, then closed down. He suggested I checked the abandoned church and monastery. That seemed more promising.

We walked into an overgrown garden, with waist-high grass and a huge mansion, featuring inner patios and its own church. A broom stood propped against the door. Someone still cleaned the church. This was far too big for Hudson's home. I discovered later it was built for an order of nuns that no longer exists. A very rich widow, Manuela Nevares de Monasterio, asked society architect Alejandro Bustillo to design it. Bustillo became famous for anachronistic, anti-modernist buildings. The monastery was founded in 1939 and closed in 1954. There may have been sixty empty rooms, but inside all we found were a few blood-sucking ticks that had stuck to us. Prestige buildings needed maintenance. It had been quickly forgotten when the funds had run out, like so many buildings in Argentina.

Rubén Ravera gave me the exact location for the other possible site of Hudson's second home. A retired art historian, Juan Carlos Lombán, had taken two years to locate Las Acacias as the Posta de Hardoy. In a privately printed book, he worked out that Hudson

could not have heard cannons firing at the battle of Caseros in 1852 (where General Rosas was defeated), as he had described, unless he was less than forty kilometres away. He also deduced that the cart journey would have taken far more than five hours, though Hudson's memory could have played tricks. So Lombán located it at kilometre 74 on Ruta 2 in the partido de Brandsen, by an abandoned YPF petrol station. It was on the main road to the seaside resort of Mar del Plata.

Miriam, our yoga-teacher friend, caught sight of a large board for Chacras de Hudson, a new country club. We parked and I walked in. A young man with dyed blond hair and an ear-ring approached. He softened when I told him who I was and why we were here and offered to drive us to Hudson's home, Las Acacias, where he happened to be living.

We followed his car along an avenue of young trees towards the brand-new but empty clubhouse and pool, the first new building erected there, in expectation of 341 more. At the moment, it was open fields, with yellow dandelions and startling red verbenas. I learned later that part of the Posta de Hardoy had been knocked down by the previous owner to make a large, shiny barn. We parked by an old, ruined dovecot that could have been there during Hudson's time. He referred to a round dovecote; this one was hexagonal, but maybe his memory let him down.

The house itself had two floors and was surrounded by an acacia wood, with some very old and large pear trees ('the oldest pear tree in Argentina', said the foreman), two or three centennial oaks and a large fig tree. This was definitely the place. I could work out the position of the Hudson's store front where gauchos would tie up their horses . . . I learned that a Fundación Hudson was restoring the old adobe-brick house, but when I later asked Rubén Ravera, he'd never heard of this foundation.

Las Acacias c1868, with a young Francisco Moreno in white suit.

The modern foreman mentioned that his boss had given him Hudson's *Allá lejos y hace tiempo* to read, but it lay by his bedside unopened. After driving round the place, we left with the owner's telephone number in Buenos Aires, while Miriam handed out a yoga invitation to the young man.

Miriam hadn't a clue who Hudson was until I briefed her along the motorway. Oddly, Hudson's name and abode had become a major selling point for buyers in a land without much recorded history. You could buy your house plot with its whiff of history surrounding Hudson, and a vague ecological intention. It looked far better undeveloped, with open fields and scattered trees.

From a tall tree, you could spot, Hudson wrote, the freshwater lake, never deeper than four feet, and famous for its pejerrey fish. There was a lake by the new clubhouse, adding authenticity to the claim that this site was Las Acacias. One photo exists of Las Acacias but I'm not sure it's exactly as it was in the days of the Hudsons (there's a tower in the background he does not mention).

Jean Leon Pallière, *La Pupería*, c1864.

At Las Acacias, Daniel, Hudson's father became a storekeeper, running that archetypal pampas store called a *pulpería*. He sold whatever passing gauchos needed. It was known as Usón's store, in an adaptation of the Hudson name. Hudson himself never called his new home a *pulpería*. He knew many of them, he admitted, from his wanderings around Azul, like the one illustrated by Leon Pallière, nearly all run by French Basques. Juan Carlos Lombán revealed that in an 1854 census in Quilmes, there were five bakers, but forty-six *pulperías*. The word's root is '*pulpo*' or octopus, perhaps dried ones sold by the store.

Luis Franco insisted on how crucial this kind of store was in exploring Hudson's Argentine roots. It's where locals bought and sold all they needed for their country lives, from ostrich feathers, to hides, knives, cheeses, yerba mate, tobacco, wine, medicines. Hanging on hooks were horses' bits, saddles, ponchos, all typical pampas stuff, but Made in England. Franco called a *pulpería* the gaucho's Mecca because it was also a bar, a social club, where gauchos from all around arrived by horse, drunk, played taba, sang on their guitars and fought. The Hudson store was no different.

Robert Cunninghame Graham wrote a sketch called 'La Pulpería' in 1898. As he knew them first-hand while in Argentina in 1870, it's worth reviving. Hudson, in a letter to him on 7 November 1898, found the story 'too brutal in its realism' and also so local that it was only for the 'initiated', like himself, for what would an English reader make of this gaucho tale: 'He would want every sentence, every clause explained.' Of course, Hudson had lived in one for roughly twenty years.

As the narrator, Cunninghame Graham tied up his horse on the *palenque* (bar), walked into a *pulpería* and ordered his red wine in a tin cup. He is dressed as a gaucho, with his *facón* or fighting knife and revolver. Indeed, he did dress as a gaucho in real-life and wore a gaucho belt. However, no Spanish words are translated. He wanted his reader to feel the strangeness. Hudson was the contrary, and avoided spraying his texts with Argentine Spanish terms. The *pulpero* – it could have been Daniel Hudson – stood behind his wooden *reja* (bars), surrounded by bottles. The house was low, squat, mud-built, surrounded by a shallow ditch, as was the Hudsons', in *tierra adentro*, that is, land claimed by the Indians. Outside were gauchos, with their toes sticking through their *botas de potro* or boots moulded to their feet from a recently killed horse, strumming a guitar and singing duels. In this store you could buy ponchos made in Leeds, sardines, raisins, bread and figs. All the trails lead to this shack like rails to a junction, wrote Cunninghame Graham. Here the narrator heard about *Martín Fierro*, for this gaucho narrative poem had returned to illiterate culture. Hudson may have objected to the story's dense realism, but he empathized with its overload of local details because he had lived it to the full.

The gauchos of Argentina played a crucial role not only for Cunninghame Graham, but Hudson too. Sir Francis Bond Head, rabidly anti-Catholic, soldier-cum-mining engineer, promoted his Romantic view of the gaucho in 1826 in his bestselling *Rapid Rides Across the Pampas*, an account of crossing the pampas four times to visit mines in the Andes. The gauchos were wild and lived in poor huts, surrounded with the litter of bones and carcasses, like an 'ill-kept dog-kennel'. Everybody lived in one room, with babies hanging in bullock hides from the roof. By four years old the children could ride, and started helping with catching and taming wild horses. They were tough thanks to their constant diet of meat and water, without milk, bread or vegetables. They were perfectly adapted to their needs. In fact, there were no luxuries, apart from a roof over their heads, dried thistles for the fire and *mate amargo* to drink. Whatever else they needed could be bought from the *pulperías*. Were they happy without luxuries? Head romantically thought so and found them courteous, always welcoming to a stranger passing by and devoted to their freedom, outside laws. They were 'delightfully independent'.

A counter-view was given by naturalist Félix de Azara, who in 1847 described the gauchos of the pampas as filthy as pigs, living in shacks without windows or doors, without furniture except for a barrel for water, drinking horns, low stools and a few pot and pans. These wild men killed another man as easily as a cow, gambled, got drunk, walked barefoot, didn't know a clock or any rules. Azara did agree with Head that horses were 'all their delight'. Their meat-only diet made them laugh at Europeans who eat 'grass'. Hudson is in the middle, neither a Romantic nor a disgusted European, but is all-accepting.

The first gaucho hut I visited on horseback in the early 1970s, near Rauch, had two or three rooms with mud floors – for the life of the gaucho had barely advanced over the hundred years since Hudson. They might be courteous and generous, but were still illiterate. And today, even though literacy has advanced, a farm hand I know well can scarcely read. I was aware of the *facón* in their belt, of their drinking and fighting, and always that tough cattle work from dawn to dusk. Coming from England, their poverty was shocking.

One last place I wanted to visit in search of traces of Hudson was in Buenos Aires. I'd heard that Jorge Casares had donated his Hudson library to the Sociedad Ornitológica del Plata. Jorge Casares came from a wealthy landowning family. He was related to the writer Adolfo Bioy Casares and to the owner of my favourite second-hand bookshop on Suipacha Street called Casares. He'd written a short Hudson biography in 1929 in a specialized bird magazine, unusually referring to him in Spanish as W. H. Hudson, and was the first person in Argentina to take him seriously as an ornithologist. It was only later that he was reinvented as Guillermo Enrique by Dr Pozzo.

After locating the webpage of the renamed bird society – it was now known as Aves Argentinas – I took a bus to the south of the city and entered the newish building on Matheu street. However, my luck was out as the librarian had been called away for a week. After a week, I returned (you usually have to do everything twice in Buenos Aires as information is unreliable and people do what they want) and was shown upstairs to the library. There had been a leak after the last of the torrential summer storms, with puddles on the floor, but the young postgrad student, standing in as librarian, showed me to a table. I asked if I had to register, or use a pencil, remembering how you read manuscripts in the British Library. He looked surprised. 'No, don't worry about that,' he said.

He had several keys with him, but none fitted the library, which turned out to be a huge oak bookcase with locked glass doors. Finally, it opened. He found a ladder and left me alone with the un-catalogued books. They had been bound in natural leather. It was the custom for landowners to bind their library books with leather from their own pedigree cattle, in this case Herefords. Casares had added his own bookplates. As I slowly pulled the books down from the shelves, I quickly realized what a complete collection it was.

After some time, I'd reached the top shelf and a large, bulging leather folder. In a funny way, it was the archivist's dream find. It was what I was looking for. It had been donated by Philip Gosse, son of Edmund, a writer friend of Hudson, and was packed with booklets and bits and pieces: two pages ripped out of one of Hudson's note-books, in execrable hand-writing, an invitation to the opening of the Hudson Memorial in Hyde Park, several numbers of RSPB booklets by Hudson, a handwritten lecture which I guessed was by Gosse him-self, given when he came over to Buenos Aires, invited by Casares . . .

Ramiro, the faux librarian, noticed my excitement and com-mented: 'Nobody has ever come here to this bookcase.' Just then a young woman came up to me, pecked me on the cheek and sat down to write, as if she knew me well and I'd always been there. I was now chatting with Ramiro, who told me he was born an orni-thologist and was studying bird behaviour, with genetics. 'The first part of your study is Hudson's territory,' I said (Hudson ignored genetics). But the student had no sense of history, nor did the bird society. He asked me, 'Did Darwin really die of Chagas's disease?' I told him what I knew, admitting that I'd published on that very subject. Somehow in the informal atmosphere I felt Hudson's pres-ence, that lament late in his life that he should never have left Argentina, that he was interesting to the scientists in England when writing on Argentine birds, not on English ones.

W. H. Hudson was made an Honorary Member the year the Sociedad Ornitológica was founded in 1916. Did they write to inform him? He was the only member from London listed in the first number of the society magazine, *El Hornero*, in 1917. But nobody there could find any correspondence or even knew where it might

W. H. HUDSON MEMORIAL

Executive Committee :
Chairman : R. B. Cunninghame Graham, Esq.

Muirhead Bone, Esq.   Viscountess Grey of Fallodon
J. M. Dent, Esq.   Halbrook Jackson, Esq.
Gerald Duckworth, Esq.   Mrs. Frank E. Lemon.
John Galsworthy, Esq.   Mrs. Reginald McKenna.
Miss L. Gardiner.   H. J. Massingham, Esq.
Edward Garnett, Esq.   Morley Roberts, Esq.
Hon. Treasurer : Hugh R. Dent,
Aldine House, Bedford Street, W.C.2

THE CHAIRMAN and COMMITTEE
request the honour of your presence at the
UNVEILING of the HUDSON MEMORIAL
by the
PRIME MINISTER
on
TUESDAY, MAY 19th, at 11 a.m.

Please see Map   R.S.V.P. to the Treasurer
on back of card   *Please bring this card with you*

be kept. Both the society and its scientific magazine were the first of their kind in South America. Hudson had a special fondness for the *hornero* ovenbird, and was quoted in several of the early scientific papers. Here in Argentina, Hudson was taken seriously as an ornithologist, rather than as a writer.

I noticed too that the poet Leopoldo Lugones had a section on Argentine birds in the December 1920 number of *El Hornero*, which I know Hudson read. He wrote poems on individual birds and must have studied Hudson.

I felt an urge to see if this unused Hudson collection could be united with the collection in the Hudson museum. I left with photocopies for which I wasn't charged, but honourably resisted the temptation to pocket my finds.

I feel Hudson's spirit was still there 140 years after he'd abandoned Buenos Aires – in the birth shack in Quilmes, the Parodi sculpture, the rented home and *pulpería* Las Acacias, and the Aves Argentinas library. But to find out more about his actual presence, I needed to return to the written word.

Chapter 2

Happy Families

So much for W. H. Hudson's meagre survival in plaques and monuments in two continents. They start to give a sense of the man but they can't quite bring him to life. I needed more details but faced the first of the many enigmas surrounding his biography and it concerns his parents. There's so little to go on. No pictures or portraits of these parents and little on his five brothers and sisters.

Professor Enrique Pedrotti, president of the Sociedad de Amigos de Hudson in Buenos Aires and a university lecturer, has attempted a family tree. In a bar in the south in Mexico street, he showed me a photo of Hudson's elder sister Carolina, who looked like any other Victorian spinster. I'd had the feeling that Professor Pedrotti was testing me, that I had to pass some exam, but then this fastidious man relaxed, bought me a toasted sandwich and shared his researches.

He showed me photocopies of hard-to-find articles. But there's not much on his Argentine days. In fact, there's only one surviving, signed photo of Hudson himself. It's a lovely image of a

Earliest photo of W. H. Hudson, 1867.

twenty-six-year-old amateur ornithologist, with a dreamy, far-off look, taken in Buenos Aires. The photo was sent to the Smithsonian and has been reproduced countless times.

Hudson's account in *Far Away and Long Ago*, 1917, is the sole source of information about his childhood, his parents, and his brothers and sisters. Hudson's best book is an elegy to the break-up of his family. With his mother's death in 1859 and two brothers away working, family life collapsed. This 'fall of the house of Hudson' is reiterative. As well as this ending of childhood bliss, Hudson's notion of how memory works is crucial: there's no order, so chronology is weak and pointless; there's no sequence and no progress. As he claims, memory serves and fools us. But then so does history. So memory and history are dreamlike and patchy, leaving more out than can be included. The same goes for biography. Outside his version of events, we are left clutching at straws: we can only go beyond him by speculating. He made it hard for biographers

for his experiences are not tied to chronology or dates. He deliberately made them 'unchangeable', eternal.

His father, Daniel Hudson, first-generation American with an
Irish-born mother, was born in Marblehead, Massachusetts, on 1 May
1804 and married Caroline Kimble, Hudson's mother, in 1825. She
came from Berwick, Maine, and was also born in 1804 on 1 October,
daughter of a pastor into a third-generation quasi-Quaker family.

When she was aged seventy-seven, Marjory Stoneman Douglas,
writer of an ecological study of the Florida Everglades, explored this
past deliberately erased by Hudson. Daniel's father had supposedly
emigrated from Clyst Hydon in Devon, England, but she found no
Hudson from Clyst Hydon in the parish record. She'd asked a parish
priest to check, but maybe he didn't look into the chapels.

Though well known as dissenters, the name Kimble or similar
did not appear on the manifest of the *Mayflower*, as Hudson had
surmised. The mother's family names of Kimble and Merriam are
well established in New England, but not much more could be
dragged out of records. The Hudsons reached America sometime
after independence, and opened a tarred rope business in the port
of Marblehead. They were 'sea-haunted', and had heard many tales
of the land that had broken away from the Spanish Empire to
become first the Provincias Unidas del Río de la Plata in 1810 and
then the Confederación Argentina or just the República Argentina.
But after arriving in Argentina, the Hudsons would remain aliens in
a Catholic culture and never really learnt Spanish.

Why on earth did Daniel and Caroline Hudson end up as cattle
and sheep farmers? The marriage certificate I saw states that they
were married in Boston on 18 August 1825, with no witnesses except
the Revd Hosea Ballou who married them. Only Caroline gave an
address – Portsmouth, New Hampshire. The rest of the certificate
is crossed out. My guess is that it was filled out in a hurry. However,

they must have brought some capital out to Argentina to buy land and animals, probably an inheritance from his family brewery.

One biographer suggests that they married against their parents' wishes; another that they had financial problems; another that a friend had spoken of the vast wealth to be made in Argentina. Morley Roberts, earliest literary friend of W. H. Hudson and first biographer, remembered something about an accident in a brewery or tuberculosis, but the fact is, as Jorge Casares noted, they wanted to become 'pilgrims' and begin again, without the baggage of family history. Obviously William's Devon-born grandfather had emigrated to begin anew, and maybe Daniel had the same desire, which the grandson later also shared. Perhaps this roaming gene is my heritage too, with my father bringing the family all the way from Mauritius to England, and myself spending a great deal of time in South America. As far as the Hudsons are concerned, all I can do is agree with Paul Theroux that the Hudsons were 'Yankees to the core'. They were among the earliest American settlers in Argentina.

What happened to the young married couple has vanished from sight, apart from some minimal documentation. They reached Argentina sometime in 1833 after a three-month journey on board the *Potomac*, a Boston-based US frigate on a world tour, after landing at the Falkland Islands. But even this date has been questioned as the first record of a ship called *Potomac*, weighing 264 tons, is two years later, 12 July 1835. Analía Fariña, who published a study of the Hudsons based on land deeds, also insisted that the first two sons were born in the United States (she's wrong, although Edwin, second son, did claim he was American in the first Argentinian census of 1869). In her versions, the Hudsons arrived in 1837, the year they bought land. If you accept her date, then they didn't delay four years in buying land. But they still delayed ten years before their first child in 1835. So another version, backed by Marjory Stoneman Douglas but without any documents, is that Daniel went out alone and called his wife to join him, so that physical separation explains the lack of children. The fact is that nobody knows.

What is established is that Daniel bought a gaucho's shack, Los Veinte-cinco Ombúes, on undulating land on 17 April 1837 from Juan Manuel de Rosas's brother-in-law, Tristán Nuño Valdéz, thirty-three kilometres from Buenos Aires by a stream called Las Conchitas. The plot measured some 336 hectares. Daniel's wife had given birth to Edwin, her second son, in the hut on 11 January 1837, prior to the purchase, so they might have been renting the same hut ever since they landed in 1833. Their first son Daniel may also have been born there.

As Fariña has shown, Daniel, the father, soon began to sell off parts of his land. Two years later, in 1839, he sold 127 hectares to Roberto Taylor, then 101 hectares to a Sr Harris. Daniel Hudson clearly needed the money. By 1846 the family could not make a living out of sheep and cattle. It seems that the whole Hudson family simply didn't understand or care about the value of money. Daniel had spent too much doing up his house, putting in a wooden floor (mud floors were usual) and a pine shingle roof, much to local amazement and the first of its kind in Argentina. They moved, around 1846, to a larger, rented house, Las Acacias, and lived there for ten years.

Then they returned to live in Los Veinte-cinco Ombúes, which remained in Hudson family hands until around 1919. I've seen a Record of Baptism in the Methodist church in Buenos Aires, where William Henry was baptized on 10 October 1841, copied into a register from the 'Family Bible'. In the year of Hudson's birth in 1841, a Quilmes judge, Manuel Gervasio López, wrote down on General Rosas's orders that the American 'Daniel Uson' (Hudson) had 300 cows and 250 sheep. Carlos Antonio Moncaut, a local historian, found another document from 1858, which records what

Daniel Hudson once ferried by cart to Buenos Aires: eight arrobas (about 200 pounds) of pig and 240 sheep hides. Hides were Argentina's principal export.

In 1859, the same year William's mother died, a judge asked for a revision of licences for selling and Don Daniel 'Huzzon' figures in a list. It's wonderful how often the spelling of the Hudson name changed from Hudson to Uson to Huzzon. His father, in a form for the Registro Estadístico Cuartel 3 in 1865, gave his name as Daniel, a North American, a widower with five children at home (Daniel the eldest son was farming in Azul). He owned a small wood, 750 sheep, three cows, five horses and thirty calves. Thus the widowed father, with the motherless family, continued to be sell drinks and food in the *pulpería*.

Then his father Daniel died at Los Veinte-cinco Ombúes in 1869 of 'general debility', according to the Methodist Church records. His age was given as seventy-four, but he was sixty-four. Luis Horacio Velázquez included in his biography of W. H. Hudson a facsimile of the land ownership but confused dates as it is clearly a map following the father's death in 1869, with one piece given to Daniel, the eldest son, and the rest to Alberto, Edwinio, Luisa, Elena and Guillermo, all given Spanish names. The five children inherited 106 hectares. In the first census in Argentina of 1869, we learn that the two sisters are in Buenos Aires, Edwin in Córdoba and Hudson, a labourer, still lives at Quilmes. Later, Daniel, eldest son, sold his separate thirteen-hectare plot in October 1877 to the Davidsons. Hudson remained at his birth home until he left for England in April 1874.

When did Hudson sell his part of the inheritance? What happened to the remaining land in 1919 when Mary Ellen, the last survivor of his direct family, died is not clear, but it seems to have also been bought by the Davidsons. Mary had preserved Hudson's room with his embalmed birds and his metal bed. She called his

room the 'Museum'. When Mary died she was living in Buenos Aires and had run a boarding house for English girls on Thames 2440.

But the main point regarding the paltry documentation concerning the immigrant family over the years of their return to the small shack is that the story told by Hudson himself in *Far Away and Long Ago* ended with the death of his mother in 1859. Not a word about living on without her, the father's long, miserable mourning, the poverty and the break-up of the family through marriage, except for glimpses in Hudson's letters to the Smithsonian and the Zoological Society, written while he was still living there.

The Davidsons, who bought the Hudson land and later donated the shack to the Hudson museum, had entered the family's life through a John Davidson (1808–93). He had lived in the Estancia Santo Domingo, visible from the Museo across the Conchitas stream. Santo Domingo had been built by the Jesuits in 1721 as a monastery. It was taken over by Dominicans when the Jesuits were expelled in 1767 from all the Spanish Crown colonies, and Davidson bought it in 1843.

I was taken there by unshaved, chatty Rubén Ravera, the Hudson museum director. The gate was open. We drove down an avenue of lime trees. The caretaker waited for us because Rubén had rung him on his mobile, and then led us to the long-abandoned Davidson house, an unusual, three-floor mock-Tudor residence on the site of the monastery. We walked around inside, inspecting the old adobe bricks, stepping on bat shit.

The garden behind the house was overgrown. Centenarian ombúes, an encroaching wood, a well, outhouses and a huge Liverpool-built cauldron, dated 1895, remained. Nearby, traditional landowner Martínez de Hoz, hated Economics minister under the Videla dictatorship, had built imposing stables for his race horses, also abandoned and invaded by slum dwellers. Shanties, a meat-freezing factory, gated communities and fields surrounded the ruins of the old Jesuit monastery. The past has been desecrated.

This lack of interest in any past was confirmed when I asked Rubén about the Capilla de los Ingleses, on land donated by Davidson in 1854. It had recently been burnt down. It's unlikely to ever be rebuilt.

In a letter to Morley Roberts of 8 April 1919, Hudson said he had spotted a short obituary in *The Times* for Sir James Mackenzie Davidson (1856–1919) and added: 'I knew him as a child – his family were unadulterated Philistines and illiterate. We rather didn't like them.' Hudson commented that one of Davidson's sons became a famous doctor – without naming him – who hardly ever visited his *estancia*. This doctor was Sir James Mackenzie-Davidson, a renowned radiologist who was knighted in 1912 and was also the King's doctor. The odd thing is that his own cardiologist shared the name of Mackenzie-Davidson. The Davidson family *estancia* featured in Hudson's best story, 'El Ombú', as the Jesuit monastery where bare-footed monks wore spurs.

To begin my inner journey into Hudson's writings, I needed to start at a deeply intimate event: his mother's death. Mother worship is a universal constant in all cultures. A point of contact is my own mother's death. Despite my complaints about not knowing who she was or about her social ambitions, and much else, I can only scrape the surface of what I owe her. Her death put me in touch with emotions that lie beyond words for I was wordless and breast-fed for eight months. Could this link be sacred? I've seen this scene in countless nativity paintings. Mother and child share this mute identity. I understand Hudson's reticence about his mother. On 5 October 1859 Caroline Hudson had died exhausted. On her death certificate the cause was 'jaundice'. She was just fifty-five years old. As a pioneer and exile, she gave birth to six children. Deeply religious, from a Calvinistic family and with perhaps a Quaker sense of 'an indwelling presence of the spirit of God', she was also a heretic within a Catholic society. Part of her Puritan tradition was a belief in direct contact with God, without reliance on priests or confession so typical of Catholicism, but her best friend, with whom she stayed for long periods, was the Methodist minister's wife in Buenos Aires.

She may have been a heretic, but her hospitality was legendary. She was from a 'furiously anti-English' family, but was also a do-gooder. After her death, Hudson found himself in a native's shack with a woman over eighty in tears, saying that though Caroline called her mother, 'she was the real mother to all of us' as she had been so kind to them. She even wet-nursed a gaucho baby at the same time as breast-feeding infant Hudson.

Intense, reverential mother-love formed Hudson's emotional core. His relationship with his mute, compassionate mother was the most powerful of his life, and no future love could replace her. A mother's love, he wrote, differs from all other kinds of love and 'burns with so clear and steady a flame that it appears like the one unchangeable thing in this earthly mutable life'.

Hudson wrote about her death in the family hut and the following devastation. It obliterated family life. Her perfect health failed her suddenly and the decline was brief. Hudson recalled his last bedside vigil in *Far Away and Long Ago*. She did not fear death, but was distressed at seeing her fourth child in precarious health and with such lawless thoughts. She merely hoped her prayers for him would prevail and they would be reunited in the afterlife.

Hudson struggled with death and afterlife. Before his mother died, he remembered a gaucho had told him he would climb on to the reed-thatched roof of his hut and spy the horizon for his dead mother's return. The gaucho said, 'And she never came, and at last I knew she was dead and that we were separated for ever – that there is no life after death.' Hudson's comment was simple: 'His story pierced me to the heart.'

Hudson hid behind Abel in *Green Mansions*: 'That is my philosophy still: prayers, austerities, good works – they avail nothing, and there is no intercession, and outside of the soul there is no forgiveness in heaven or earth.' No God, no return, for death is final. His lack of Christian faith contrasted with his mother's and was the source of his lifelong melancholia at a vanishing world. After she died in 1859, everything else was loss, including home. The death of his mother and her internalization as sacred memory made him a nomad, a roamer, for there was no substitute for mother.

Hudson shared his mother's reticence, but there's a curious scene in his novel *A Crystal Age* that reveals his wish to in some way commune with his mother after she had died. Smith, the protagonist, is received by the dying mother of the future utopian society. She asks him to place his hand, like a healer, on her head to relieve her dreadful pain. He does. She remains silent and moans. Smith then 'whispers', asking her if she wants to say something to him. She tells him not to be afraid; she will not die.

It's clear to me that Hudson had re-enacted in fiction a bedroom scene with his dying mother, with the innuendo that Hudson was chosen as her favourite (something all children want to think). When in the novel Smith listened to a group sing, he was suddenly reminded of 'my beloved mother, whose early death was my first great grief in boyhood. All the songs I heard her sing came back to me, ringing in my mind with a wonderful joy, but ever ending in a strange, funereal sadness'. The vibrating air left him in a state of 'exquisite bliss and pain'. His mother's singing of lullabies and hymns united them in his trembling mind, beyond death. For Hudson, only birdsong and wildflowers would touch the same depths.

His adolescent mind was torn between religious faith and scientific fact, with the beauty of wildflowers voicing his silent mother. Sharing this natural harmony created a 'secret bond' that differentiated him from his siblings. He wouldn't rejoin his mother in the afterlife, but wildflowers spoke to Hudson in primal mother-speak. He shared a 'spiritual kinship' with her, expressed in a mutual love of pampas wildflowers, little voiceless messengers, 'which I never see in England'. To see them again, he wrote in 1916, would be to commune with her again.

Once out riding he discovered a 'patch of scarlet verbenas in full bloom, the creeping plants covering an area of several yards, with a moist, green sward'. These sweet-smelling, creeping verbenas are common on the pampas and escape being nibbled by sheep. Hudson wrote some sixty years later, as if time hadn't passed, that 'I would throw myself from my pony with a cry of joy to lie on the turf among them and feast my sight on their brilliant colour.'

His evocation of the delicate, yellow *macachín*, the first wildflower to blossom on the pampas, is another secret elegy to his

EVENING PRIMROSE

mother. Hudson did not care for cultivated roses or well-trimmed gardens, only wildflowers. He would eat this *macachín*, with its acid taste, and dig out the bulbs and eat them too, tasting of sugary water. He was dazzled by wildflowers swaying in the wind. Gorse had a narcotic effect on him; making him languid, ready to 'swoon in that heavenly incense'.

The wildflower that stirred Hudson most was the evening primrose. His closing chapter of *Idle Days in Patagonia* hovers round this 'pale-flowered alien' in England. To stop and thrust his nose into this weed is his 'kind of religious performance'. For the evening primrose was abundant in the pampas. At the moment of writing, holding his fountain pen, Hudson imagined he was holding an evening primrose. It linked him with his past and 'summons vanished scenes to my mind'. Back there, he knew this weed by its Spanish name, *Santiago de la Noche*. To him, its scent is a miracle where time and space become 'annihilated and the past is now'. Despite being in London, writing at his desk, he was also sleeping under the stars, waking up to the subtle scent of the evening primrose. What he didn't confess was that his mother was there too.

The root of his vivid, emotional experience is simply grief. And the most abiding grief was for his dead mother. Scent best recreates this 'irrecoverable past'. *Idle Days in Patagonia* closes with a black-and-white illustration called 'Wakening at Dawn' by his Shoreham artist friend Alfred Hartley: Hudson is asleep by a patch of evening primrose, his horse wrenching grass nearby.

Daniel Hudson survived his wife by nine years to die in 1869 in the same hut as his wife. From Hudson's account, while his mother Caroline gave unconditional love and religious awe, his older brother Edwin gave him the challenge of thinking for himself. Was Daniel an ineffectual father? It's hard to decide on his paternal role. He was honest and sincere, but doesn't seem to have had much to do with the children. He had to learn farming cattle the hard way as he came from a brewing not a farming family. He had also to learn Spanish from scratch, aged twenty-eight.

He obviously remained a *gringo*, or foreigner, as the English surnames of the two men to whom he sold land proclaims. Gauchos particularly mocked *gringos* (usually Italians who knew nothing about horse-life). There's a scene in the narrative poem *Martín Fierro* where an Italian *gringo* is taunted from how he talked to his shivering in the cold and not spotting ostriches. He's only fit 'to live amongst homosexuals' '*pa vivir entre maricas*'. Daniel could have been that *gringo*. The word *gringo* itself points to linguistic confusion as it derives from *griego* and *hablar griego*, or Greek to me. Furthermore, Daniel did not drink, never swore, was impassive and fair and never punished his children, qualities that made him appear weak and different in a macho society where the exact opposite ruled. Hudson labelled him a humble-minded and literal man whose blind spot was his lack of understanding poetry.

One incident stands out in Hudson's patchy but vivid memory concerning the defeat of Rosas by General Urquiza at the battle of Caseros in 1852 when he was eleven. Streams of horsemen rode by in their scarlet Rosas uniforms and odd three-cornered hats, stealing horses as they passed. His father was warned that his family was in danger as these defeated soldiers were without officers. So he hid his horses in the wood behind the houses at Las Acacias. He laughed off the family's fears and said 'they won't hurt us'. In the hot mid-afternoon (it was early February) about ten men turned up and demanded horses. His father met them and, smiling, said he had none. Everyone hid indoors, including an old friend with his sword drawn. His father continued to smile when the thugs refused to believe him. After drinking cold water, a man violently rattled his hilt as if he would kill Daniel. The father still smiled as if this was a joke. Hudson knew his father wasn't acting. He just didn't have fear, not even of lightning during the extraordinary violent summer storms.

Hudson characterized his father as ordinary, with not a thought of getting rich. He inherited – and suffered – this paternal scorn for money. He was also more neighbourly than most of the Argentines in the vicinity. He was civic-minded before civic society existed, formed in the values of the United States. He also had a disastrous child-like trust in his business partners that led to his financial doom.

In a land where daily violence and sudden death were the norm, Daniel stood out as a natural pacifist. Hudson only recalled a single violent moment when they found an owl was killing his doves. He refused to slaughter animals. A gaucho view of this exiled puritan would have been pitiless and his qualities would have become terrible defects. Hudson recalled his father's rage when a visitor suddenly raised his rifle and shot a passing swallow. I see these aspects of Hudson in his outcast father. However, Hudson never mentioned his father's death or his family's bereavement. He lived on among his father's things in the shack for another five years, the last to leave the pampas home in 1874.

I'm not sure I have grasped the peculiarities of either the mother or father. I could fill in the blanks by basing their characteristics on stereotypical New England puritans in exile in Argentina, without touching their quiddity. Ordinariness is difficult to portray except as stereotypes. I don't even know what they looked like, let alone what they thought. They were clean-living and devotional and hard-working, without much time to be idle or think. At the same time, they were happy in their family enclave with a hostile world outside, a kind of Swiss Family Robinson, or at least they did not have the time to be unhappy.

Their first son Daniel Augustus was born on 18 June 1835 and died in 1889, aged fifty-four. He was born two years before his parents bought Los Veinte-cinco Ombúes, and two years after his parents landed at Buenos Aires. He was only baptized four years later. Those early years are a blank. He became a smallholder, one time renting land near Azul. He assumed his father's mantle of sheep and cattle farmer. He had inherited the largest part of Los Veinte-cinco Ombúes, but finally sold it in 1877 to the Davidsons. He had five children. W. H. Hudson went and worked with him in Azul after completing military service together, as we'll see. But Hudson only referred to his eldest brother in a few asides. He recalled a Basque peasant singer at his Azul ranch in the wilderness whose song about a girl saying goodbye to nature had haunted him. He also admits his elder brother had a musical knowledge 'which I have never possessed'. Daniel farmed on land near the 'frontier' with the Indians, which was cheaper to rent as it was dangerous.

Hudson related to Morley Roberts an incident that his brother lived when a party of Indians rode up and broke into his farm – he was luckily absent. They seemed to object to paper. All they did was to shred all the paper, so that when he returned it looked like a small snowstorm had taken place. Another incident concerns Hudson, on the way to visit his brother at Azul, being charged down by two Indians. He stopped and drew his revolver, but they passed galloping and shouted in Spanish, 'Friend, your horses are tired!'

He and his eldest brother travelled around in a gang of men, staying in each other's huts and telling stories and smoking. One poor gaucho kept them up all night talking about everything. 'He was the greatest genius I ever saw,' said Hudson much later in London. But he didn't stay long on the frontier with this brother as he felt too exposed.

By far the most influential brother, the leader of the family pack, was Edwin Andrew Hudson, the second son born in 1837. As Hudson remembers with awe, Edwin looked after the guns, went hunting and exposed the passing tutor Father O'Keefe as a fraud who knew no maths. The subject was his strong point. He refused to become a sheep farmer, resisted military service and insisted on being educated abroad in the United States. From around 1852 he spent five years in Boston with his uncle John Hudson, studying land surveying, but no biographers have provided details. He brought back the fatal copy of Darwin from the United States that changed Hudson's life.

Hudson's chapter 'Brothers' in *Far Away and Long Ago* deals with Edwin as tyrant. Edwin was four years older and crucial to Hudson's development as he was more virile than their father. But he was not close to Hudson's heart. He brought stories home of violent men like Jack the Killer, an Englishman turned gaucho, who fought and drank and gambled like any gaucho but he tended to kill not wound opponents, contrary to the gaucho way. Jack hated being mocked as a *gringo*. He was Edwin's hero. And Edwin learnt how to fight thanks to Jack.

One day Edwin invited Hudson to mock fight with knives, gaucho style with a jacket over the left arm as a shield. But it went too

far and Hudson received a deep gash from Edwin's knife that survived a lifetime: 'It was a deep long cut, and the scar has remained
to this day, so that I can never wash in the morning without seeing
it and remembering that old fight with knives.' To young Hudson's
credit, he didn't betray his brother's knife wound to their mother.

Another anecdote from Hudson has Edwin reading a novel,
when their mother questioned whether he should be reading such
romantic stuff and not furthering his studies. Instead of exploding,
as Hudson noted, Edwin just said that it would be the last novel he
would ever read. That he became a tough, practical man was the
consequence. Reading novels in that Quaker house was not deemed
healthy. It reminded me of Edmund Gosse's non-conformist
upbringing where 'no fiction of any kind' was admitted. Everything
had to be literal and true. Hudson himself would remain all his life
ambivalent about reading fiction.

But it's when Hudson evoked Edwin as 'touchy' and
'confoundedly independent', that we see family resemblance. This
tough older brother Edwin also initiated Hudson into a love for
writing and dreaming. Edwin was the storyteller. Hudson recalled
that 'he had an extraordinary talent for inventing stories, mostly
of wars and wild adventures of fighting'. Once the boys had been
put to bed in their shared room, in the dark, he would begin 'one
of his wonderful tales and go on for hours, we all wide awake,
listening in breathless silence'. Could storytelling be better evoked?
Edwin then abruptly decided to end his stories with a 'no more
stories' and took to reading on his own, and told his siblings to
hold their tongues or clear out of the room. All storytelling is oral
in origin, but usually bedtime stories are told by mother. It's an
odd coincidence that my younger brother and I began an on-going
storytelling, lights out on our bunks in our first house in London
in the 1950s. We'd bag characters and actions and thread them into

long stories. I marvel at this oral inventiveness for as we jointly told these stories the characters would spring to life and we saw and heard them. No parent was involved, as with the Hudsons. We'd turned our reading of comic books with Kit Carson, Buck Jones, Tom Sawyer and Davy Crockett and become our own heroes.

Edwin also led a memorable and gory childhood adventure. Moved by epic battles – he had been reading ancient history – he proposed to slaughter *escuerzos*, native toads that abounded in a dry gulch near the Las Acacias house that flooded when it rained. This fearful creature swallowed its prey alive. The boys emptied a drinking trough and floated off, capsizing and spiking toads with sharpened bamboos. They slaughtered some eighty and arrived back home soaked and cold. When the young Hudson speared a toad with his javelin, he felt such disgust at this gratuitous cruelty that he still, in 1916, over sixty years later, experienced it as a nightmare. Hudson evoked toads from his copious notes in 1892 as hideous beyond description with brilliant green skin, bright yellow lips and about the size of a man's fist. The pale gold-colour eyes can be elevated or depressed. It recalled so many nights when we were staying at my wife's estancia and these toads would sing like a wind instrument, heard miles away. So many times after storms when we were stuck in the country with the mud roads impassable, we would hear this frog and toad din all night, the escuerzos the oddest. It would keep us awake.

Hudson noted also how these toads swell and bite. Once a gaucho friend had to cut one off his hands (it locks itself on its prey), just in time before the poison took effect. One summer, he wrote, two horses were found dead near his home, with these vicious toads still hanging to the cadavers.

Hugo Backhouse, born in 1890, who ran off aged sixteen to become a gaucho and ended up as captain of the Argentine Polo team, evoked the *escuerzo* as one of the dangers of gaucho life. It blisters skin if touched and once it bites a finger it's impossible to make it let go. It showed no fear and attacked at the least provocation. It swells in size when irritated to the point of actually bursting.

Edwin's return home to Los Veinte-cinco Ombúes just before their mother's death in 1859 was 'an event of the greatest importance' in Hudson's life. He reappeared as another person, so greatly had he changed. Edwin had left as a sunburnt wild gaucho, with dark piercing eyes and long black hair, more Pampas Indian than white. Now, Edwin had grown a beard and moustache like any civilized grown-up. Hudson too would grow a beard, like a good Victorian, and never shave it off. But Edwin's character hadn't changed. He interrogated Hudson and mocked his belief in their mother's family creed.

Hudson knew nothing about Charles Darwin, so Edwin lent him *On the Origin of Species*. After a first reading, Hudson told Edwin that Darwin had simply disproved his own theory. Natural selection had not produced any new species. Edwin insisted on his re-reading Darwin, this time with an open mind and not prejudiced by religion. By now Hudson had recovered from a serious illness and was back to hunting and riding, but over the months, thanks to Edwin, 'insensibly and inevitably I had become an evolutionist'. Hudson, in his own words, 'had come out of the contest a loser' in the battle between his mother's faith and Darwinian science.

A detail that confirms this defeat of mother was that in 1860 Hudson became a probationer at the Methodist Church, but was never received, unlike this younger brother. He had given up 'religion'. This lending of Darwin's *Origin*, the greatest book of the nineteenth century, and then being forced to re-read it, was a fatal touch that made this tough elder brother's influence far weightier than that of his father.

Edwin, then, administered the break with Hudson's moth-er-shared, intense religiosity. He himself had shed intellectually all belief in a Christian creed and boasted about it. But he never told their mother, even though she knew. Hudson argued that his elder brother Edwin could do this because he had such perfect health, while he, Hudson, had confronted imminent death with his chronic heart weakness. Yet Hudson outlived his brother by thirty-three years.

Edwin Hudson left the family home and worked as an engineer and a land surveyor around Córdoba. He surveyed the railways, the mountains, woods and watercourses of Río Cuarto in 1869, as land was opened up after the expulsion of the Ranquel Indians. He was allocated an armed escort in this dangerous, frontier territory. His remuneration for this important work, as Ana Inés Ferreyra noted, was a gift of 5,000 hectares of land. He also surveyed Río Tercero in 1872. Family lore, passed on by Violeta Shinya, has him meeting the North American railway magnate William Wheelright and surveying the line from Salta to the Bolivian border, now part of the tourist Tren a las Nubes line.

Edwin went on to become Córdoba's Municipal Engineer in 1875. He instigated anti-flood barriers and fixed bridges to become the highest paid employee in the Municipalidad. In 1876 he organized the water flow to the Fountain on the Plaza San Martín in Córdoba; his name can be seen carved on the fountain's side in Carrara marble as 'E. A. Hudson'. He retired in December 1879 and died at the age of fifty-one. He was buried in the Córdoba Anglican Church on 16 January 1889.

In a 1904 letter to A. R. Wallace, Hudson refers to this brother who had been living in Córdoba, who was in a good position and who could have got whatever Wallace sought, 'but alas! He is no more'.

Edwin was the most eminent of the Hudsons and many docu-ments remain to be unearthed. In *Birds of La Plata* it was Edwin's reaction to the novel *The Purple Land that England Lost* that struck Hudson as crucial, harking back to Edwin's role as educator. Edwin had written a letter in Spanish, which Hudson translated thus:

> Why are you staying on in England, and what can you do there? I have looked at your romance and find it not unreadable, but this you know is not your line – the one thing you are best fitted to do. Come back to your own country and come to me here in Cordova [*sic*]. These woods and sierras and rivers have a more plentiful and interesting bird life than that of the pampas and Patagonia . . .

Hudson read that letter in Spanish with a pang, thinking that Edwin was right, but that his message was too late. Hudson had made his choice to 'remain for the rest of my life in this country of my ancestors, which has become mine'. When he re-read that letter, two years before dying, the pang returned and he realized that he had probably chosen the 'wrong road of the two then open to me'.

Hudson's younger brother Albert Merriam was the person he felt closest to in Argentina. Born on 30 August 1843, he was two years younger. It was Merriam who bade him farewell on the *Ebro* in 1874. And it was to Albert Merriam that he wrote a long letter about his sea journey to England, now held in the Hudson museum.

While still at home at Los Veinte-cinco Ombúes, Albert Merriam became deeply religious. He was accepted into the Methodist Church. He had read Herbert Spencer, and other philosophers and evolutionists (so three Hudson brothers confronted Darwin), and had concluded that the materialists had won, that mind was a function of the brain that rots. However, he knew more, as Hudson underlined, and was sure that we had a soul. Hudson called him a mystic. Albert Merriam's way of combining religious faith and

Albert Merriam Hudson.

evolutionary theory was close to that of Hudson. He taught English at Buenos Aires's prestigious Colegio Nacional and was a Methodist pastor. By 1883 he defined himself as 'argentino'. He was twice married, with four children. A son Huberto became a Contador Público Nacional and corresponded with Morley Roberts over his biography of his uncle W. H. Hudson. Like his brothers Edwin and Daniel, Albert Merriam died fairly young, in 1893 aged fifty.

Mary Ellen, the youngest sister, born on 30 November 1846, survived the longest and died in 1919. All the Hudson birthdates and baptism dates and marriages are registered in the Methodist Episcopal church in Buenos Aires on San Martín, which traveller Richard Burton found 'American, hideous', but there's nothing about their deaths. Mary Ellen's life was extraordinary. She married an alcoholic, Thomas Denholm from Scotland, in 1870, 'in the camp near Buenos Aires', as the Methodist records puts it. W. H. Hudson was a witness. After the birth of eight children, Denholm then abandoned her.

She moved to Córdoba to be under brother Edwin's protection. Mary Ellen lost her only son to a drowning accident and six daughters to diphtheria. There's little to go on, but what's clear is that only one child, her youngest, Laura, survived. Hudson told his RSPB friend Mrs Hubbard about a nephew, due to be married and to visit England. He had been sent to Patagonia by the railway company he worked for, became run-down and died – 'Sad tidings from my sister' – but it could have been the son of another brother.

Mary Ellen's sole surviving daughter Laura married Yoshio Shinya in 1907. Here's another story. He was the first Japanese immigrant in Argentina, arriving on the frigate *Presidente Sarmiento* in 1900 as helper to the cook. After Laura's death in 1915, Mary Ellen, with Shinya, her son-in-law, brought up her sole grandchild Violeta Shinya, who became the first director of the Hudson museum.

Sadly, Hudson destroyed the 'frequent' letters he received from Mary Ellen. Only snippets of their correspondence have survived. For example, he had posted her his articles (one on 'Moths' from the *New Statesman*, for example) and sent *Far Away and Long Ago* with a dedication 'to my dearest sister Maria Elena' (note the Spanish), dated New Year's Day 1919 in his hand (it's held at the Hudson Museo), but she died before it reached her. Or, again, in a letter to the Ranee of Sarawak in 1915, Hudson mentions 'my beloved sister in Cordoba' who congratulated him on having such a democratic friend as the Ranee, as English women in Argentina were snobs.

Hudson announced the death of this sister Mary Ellen in another letter dated 30 October 1919: 'so long severed, so far apart – some seven thousand miles of sea – and yet I seemed to be going through life with her hand in mine. And now it has slipped out of mine and I feel a certain sense of loneliness.' Mary Ellen was in touch – held his hand – with William all through life, despite the distances. She had retained his belongings in the family *rancho* in case he ever returned.

His elder sister Carolina Louise was born in 1839, died unmarried in 1902, and barely enters Hudson's memories. She had been a nanny with President Mitre's family.

All five of Hudson's siblings remained Argentines. In 1894 Hudson confessed, in a bleak moment, to Cunninghame Graham that, 'I had brothers and sisters and other relations out there, but whether they are living or not I cannot say.' Whether he could or couldn't say, he destroyed the letters from them. Maybe he wanted to be unique and not belong to a family. His journey towards becoming English was his own, not a family inheritance.

Chapter 3

Ombúes, Gauchos, Indios, Urquiza, Rosas

The avenue of twenty-five ombúes stood out on the slightly higher ground of the Hudson home. The single ombú I saw on my first visit to Los Veinte-cinco Ombúes in 1987 is still standing today. The laconic horseman who told me back then that the tree was over a hundred years old wasn't exaggerating: Hudson himself calculated those 'trees' were already over a hundred years old, which means the sole survivor must be over 200 now. The largest became a vast treehouse for the four boys.

According to communist critic Luis Franco, the ombú is a 'sacred' tree, indigenous to the pampas, home to countless animals, birds and insects, offering delicious shade in sweltering summers with its lush evergreen leaves. Its Guaraní root, *Imboú*, means 'tree that attracts the rain'. Hilario Ascasubi, gauchesque poet, in 1872 (while Hudson was still in Argentina) wrote that the ombú offers its 'gigantic shade' and is a 'beneficent tree'. It's a 'romance in itself',

Ombú seeds.

added Hudson. Useless as firewood or for furniture, the spongy bark crumbles when it dries. Its leaves were used as medicine. Richard Burton exaggeratedly evoked the ombú in 1869 as 'thick-headed, gouty-footed'. You would never know they weren't trees. In the margins of a George Santayana book, Hudson scribbled that his ideal tree was 'probably an ombú'.

In an attack of nostalgia, and provoked by his new friend Cunninghame Graham, his sole writer friend familiar with Argentina, Hudson published his short story 'El Ombú' in 1902. Edward Garnett deemed it his masterpiece. The first edition carried an epigraph from a poem by Luis L. Domínguez in Spanish, which in later editions he cut out. The poem stated that if the most prominent trait of Brazil is sun and Peru its mines, Argentina's is the ombú of the pampas. Hudson could recite this Domínguez poem by heart. When the gaucho Martín Fierro, in what's become the national epic of Argentine

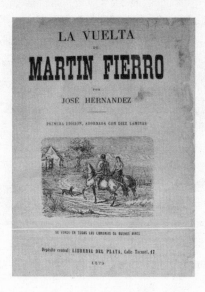

Cover of *La Vuelta De Martín Fierro*, 1879, showing the
Azul Mountains in the background.

literature *Martín Fierro*, 1872, finally managed to escape being tortured
as a prisoner of the Pampas Indians, he crossed the immense green
plain with only sky and horizon, until he glimpsed a 'sierra' (probably
the sierra of Azul): 'at last we stepped on land / where the ombú
grows' – that solitary tree signalled *home*, as it did for Hudson too. Or
the end of home.

By writing this tale, Hudson returned to his past. Cunninghame
Graham always referred to his friend as a 'latter-day Argentine'.
El Ombú – his short-story collection that became *Tales of the Pampas*
in the United States and *South American Sketches* in England, though
Hudson had preferred *Tales from the Argentine* – is dedicated to R. B.
Cunninghame Graham, 'my friend', with a phrase in unaccented
Spanish: (*'Singularisimo escritor ingles'*), 'who has lived with and
knows (even to the marrow as they would themselves say) the horse-
men of the Pampas, and who alone of European writers has

rendered something of the vanishing colour of that remote life'. He differentiated Cunninghame Graham from all other writers; only he could grasp Hudson, only he knew the gaucho proverb *'hasta el tuétáno'* ('even to the marrow') behind his English. This deep knowledge could only be imbibed from birth and Hudson lived it for thirty-two years until he left on the *Ebro*.

According to his appendix to this 'true story', Hudson had found undated notes with pampas dust inside from the great dust storm of 1868. He had listened to old gaucho Nicandro's story in 1868, while gauchos were still anarchistic, hating authority and showing no patriotism (made evident later in *Martín Fierro*, where the persecuted gauchos Fierro and Cruz preferred to live with the wild Indians than enter civil society). Luis Franco thought the gaucho, between 1855 and 1875, was possibly the most intensely individualistic type known to man. In England, only Cunninghame Graham would appreciate this gaucho tale of doom, as some points may 'strike the English reader as very strange and almost incredible'. English readers, wrote Cunninghame Graham in an introduction, miss something. To appreciate Hudson, he continued, 'they must have looked out on that grassy sea . . . They must have ridden in the scorching sun for hours.' Of course, he was wrong, for Garnett loved it.

The story echoes Poe's 'The Fall of the House of Usher' and evokes a ruined house, for Hudson has his narrator state that sorrow comes upon the house on whose roof the shadow of the ombú falls. Even if you sit in the shade you become 'crazed'. Hudson altered the usual gaucho association with this tree to make his story a parable about the impossibility of happiness. In his story he referred to a specific ombú near to his rented home Las Acacias that was so enormous that five people with outstretched arms could just encircle it. This ombú belonged to Doña Lucia del Ombú, a venerable grandmother with attractive granddaughters who lured men from

all around during cattle-branding. Her ancient ombú was a land-mark and a surname.

Nicandro, an aged gaucho, tells a series of grim, surrealist stories. A twenty-seven-year-old Hudson, working like a gaucho himself, sat and listened. What's crucial to remember is that Nicandro was illiterate and his oral tales are local history. Hudson listened to countless gauchos voice their laconic, even aphoristic wisdom, what Luis Franco called 'illiterate philosophy'. From them he learnt how to convey 'voice': his later writing carried this oral gaucho tradition onwards by stressing his own voice. I hear him speak directly to me as I read. I overhear him translating his gaucho narrator from those evenings when there was nothing else to do but sip *mate* round the kitchen fire. He was constantly worried that 'the flavour of gaucho talk is lost in translation'. He had to recreate it. But no doubt the human voice is the primal natural music: 'and, best of all, there is the human voice . . . speech, the sweet music of it, is infinitely more to us than song and the sound of all our musical instruments'.

Prilidiano Pueyrredón *Rest in the Country*, featuring an ombú, c1868.

Nicandro the gaucho would never have grasped his far-off English audience. England was so unimaginable that in the narrative poem *Martín Fierro* 'Inglaterra' became *'Inca la perra'*, but the pun is lost in the English translation, 'Inca bitch'. Still today when you travel from Buenos Aires to isolated farms around Azul, you time-travel back to Hudson's oral world. The *peones*, the labourers who work the land, are to all intents and purposes illiterate. These narrated events were not witnessed by Hudson, but by a gaucho called Nicandro. This tale does not reflect his own experiences. What matters is that Hudson fixes these oral tales in written prose.

He praised his gaucho narrator Nicandro's memory, 'since he could remember and properly narrate the life of every person he had known in his native place'. Memory, so deep and constant in his own life, furnished him with most of his fiction, set in the South America he had abandoned in 1874. His solitariness ensured that memory remained precise and emotional, as if talking too much might lose it. In 1922 he described gauchos as 'these lovers of brief phrases', a characteristic of himself. Hudson could recall and whistle the exact songs of some 154 Argentine birds. He never forgot, like Nicandro.

Over the years that I have visited and lived in Argentina since 1970, I've noticed how men and women who live isolated lives in the pampas retain acute memories. Less happens to them. They are not overloaded as many of us are with urban rubbish and intellectualizing minds. Their senses are more alert. A *peón* recalls exactly what I said four or fourteen years before, where we stood talking, what clothes I wore. I have forgotten. For the *peones* like Diego Deris, a day is made up of watching me walk along a wire fence, eating a choripan, a visit to us on horseback and much listening. The little details stand out. Maybe their sharp vision leads to sharp memories. I tend to recall only what I've jotted down in my notebook, while he catches the whole scene. He demands less. A unique countryman's cunning is the result. The *peón* will never tell you something doesn't work until you ask him about it. His mind, freed of books and print culture, thrives on trying to outwit you. It made me wonder if there were advantages in book learning when it dulls the senses and only leads to memories of reading. As Borges once quipped, he remembered more what he read about than what he lived. I've learnt to respect the *peón* because his mind is impenetrable, beyond my language, and his senses are the more acute. Culture hasn't arrived like a curtain shutting in a room.

He listens and thinks for himself. Borges's Uruguayan gaucho Funes had such a prodigious memory that he had to live in a dark room in order not to remember more. He had chosen a gaucho to be burdened by memories. Hudson, Nicandro, Funes and this *peón* faced oblivion with awe-inspiring memories. The story 'El Ombú' is a string of vivid memories told by a gaucho.

The first memory concerns the original owner of the cursed shack under the ombú, a Don Santos Ugarte, with his white skin, foul temper and air of authority. He had illegitimate children everywhere. Nicandro's memory reaches back to 1806, before Argentine independence. Santos Ugarte rode off to a monastery (the abandoned Santo Domingo next to Hudson's home, belonging to the Davidsons) with gifts for the monks that include a side of fat beef, a suckling-pig, some lambs, a few fat turkeys, partridges, armadillos, the breast of a fat ostrich and dozens of ostrich eggs. How to explain to an English reader what a pampas armadillo is? How tasty its flesh cooked in body armour is? On an isolated farm in Corrientes, I watched with some disgust an armadillo being cooked on an open grill (parilla), but once ready its flesh was white and tasty. Do tastes mean anything to someone who has never tasted it? It's the old philosophical conundrum that if you don't already know Cheddar cheese, you won't know the word and how it tastes. It's the actual savours released in a mouthful that count, not your skill in comparing the experience to eating chicken or rabbit. Santos Ugarte travelled with a gaucho's cornucopia – not a vegetable in sight and no carbohydrates.

Santos Ugarte and his gauchos crossed the Samborombón river and unexpectedly bumped into the English invaders in 1806, on their way to conquer Buenos Aires while it was still a Spanish colony. After governing the city for forty-five days, the English were ejected by a popular militia and in 1807 they were defeated a second time when they tried to reinvade, a foretaste of the beginning of the rebellious break with mother Spain in 1810.

Hudson recalls a detail: the English gave away blankets used to cross a muddy river. That's all. No overview, no understanding of historical forces, no interpretation. Just a vivid anecdote, for Hudson has reduced history and his own life to anecdotes. He thinks in anecdotes, refusing to interpret. His model is not only the peasant and ballad singer who tell stories, but also the Bible and its parables. The English soldiers laughed at the way the locals grabbed blankets, but when they saw the monks were barefooted with spurs on (even monks had to ride to exist), they cried out: 'We are sorry, good brothers, that we have not boots as well as blankets to give you.'

While writing the story Hudson checked the archives at the Royal Hospital, Chelsea, and found it confirmed in the proceedings of General Whitelocke's court-martial in London 1808. This same failed invasion titled Hudson's first published book, *The Purple Land that England Lost*, 1885. In the original first chapter (later edited and tucked away as a succinct appendix), Hudson penned a thumbnail history of Uruguay, the purple or blood-soaked land, from Magellan up to the English invasions, which 'burst like a terrible thunderstorm'. Nobody in Buenos Aires had forgotten these 1806–7 invasions, even if the English had. When I read patriotic Darwin on his travels in Argentina, barely twenty years after, there was not a word about that defeat.

Nicandro's yarn has Santos Ugarte involved in a game of El Pato (meaning duck), where a large duck or turkey is sewn up with four tough handles and horsemen vie to rip it away from each other as a prize. Rosas later banned this game. Hudson's appendix to the story explains this 'ball' game, played while the English Army invaded, as 'very strange and incredible' to an English reader. But years later in 1953 Juan Perón revived the game, substituted the duck for a large leather ball with six handles and decided to make El Pato the National Sport (not football or tennis). Hudson touched on an attitude that he too shared, 'for the gaucho is, or was, absolutely devoid of the sentiment of patriotism and regarded all rulers, all in authority from the highest to the lowest, as his chief enemies'. In *Martín Fierro*, 1872, the two gauchos deserted the Army, had no grasp of why they were fighting and remained anarchistic individuals. There was just no society to be included in. Their lives were so isolated, so far from information and news, that only with the recent arrival of satellite television are country dwellers on the pampas involved in social life. It's not been helped by the inhabitants of Buenos Aires just ignoring them.

In Nicandro's tale, the commander of Valerio, a *rancho* dweller, was a cruel Col. Barboza (a neighbour mentioned in *Far Away and Long Ago*) who had him flayed alive for voicing his companions' complaints. This same Barboza reappeared in this tale as a general who had a would-be assassin killed, but felt he'd been cursed. He called in a *curandero* – witch doctors are still consulted today, something that fascinated V. S. Naipaul – who said he would only be cured if placed bodily inside a living bull. As the sliced-open bull bellowed and soldiers crowded around, the General leapt out naked, dripping red in bull's hot blood and ran into the plains until he dropped dead.

Old man Hudson remembered that once at a dance in a gaucho's house, he entered into a discussion with a group of men as to the merits of the frontier and desert life. It taught a man to rely on himself, to be quick to deal with danger, to take proper care of his horses and, above all, it made him a man. A ruffian named Bruno López defended civilized life as frontier life was too lonely, no friends, no wife, no fireside. He especially hated the partridges calling to each other – Hudson always has some bird at the core of his memory tales. Did this partridge call emphasize his loneliness? Was its song too like a woman singing? This common bird, *Rufous tinamu*, known as a *perdiz grande*, has a very beautiful voice, with an evening call composed of two long clear notes, followed by a tri-syllabic one. The ruffian tried to whistle this partridge call and failed, so that all the gathered men laughed. For Hudson this bird, mistakenly called a partridge by the Spaniards, represented the 'desert pampas'. Gauchos would say they were off to 'where the partridge sings', meaning the deserted or Indian pampas. The almost identical *perdiz común* will wait in tufts of grass until the last minute before whirring loudly away in what Hudson called 'a surprising noise and violence', comparing its din to a vehicle 'driven at great speed over a stony road'. On a stroll on the pampas at my wife's estancia, I'd be threading my way through tufts of pampas grass, in my usual state of day-dreaming and observing, when, suddenly, at my feet, a *perdiz* would fly off in a curving arc to another hiding place. The noise of this bird's whirring wings always startles me. When I used to ride, it was worse as my horse was as jolted as I was.

By now any reader will grasp how removed Hudson's world was from Victorian London. The details that his stories compact of this 'semi-barbarous life of the plains' – raw mare's flesh, barefoot monks with spurs, witch doctors, random acts of cruelty, relentless bad luck – would strike the English reader as too unfamiliar. What helped Hudson's portrayal of the gaucho was that these gauchos lived in the present moment, without past or future, that is, out of historical time. History happened in Victorian London.

S. J. Looker, a critic, saw Hudson as too unfamiliar. He was 'partly so difficult to understand because he came to England from a far, strange country . . . He never lost the touch of strangeness, as of a foreigner who views another land than his own.' Looker could not conceive that 'El Ombú' was true, though couched in fiction.

He mocked Hudson's 'maidenish reserve', but it was Hudson who had experienced the gory details of lawless life and who noted a Victorian or 'modern squeamishness' that had replaced Chaucer's ranker zest for life.

Cunninghame Graham touched on an extraordinary insight. That when Hudson wrote this story 'El Ombú', 'he must have thought in Spanish'. Hudson's oddity is linguistic as well as cultural. To appreciate Hudson is not only about galloping bareback, but 'one must know Spanish', as he did. Cunninghame Graham noted in Hudson's final work, A Hind in Richmond Park, 1922, that he wrote 'thinking in his half-forgotten, bi-maternal tongue'. All his short stories recreate what had been recited to him in Spanish. He lived in a Spanish-speaking environment for the first thirty-two years of his life. Spanish, with American English spoken at home and with certain friends, was his emotional language from birth, through teenager years into outback adulthood. Language becomes dormant if you don't speak it, but, like bike riding, is never forgotten. At one moment in 'El Ombú' we read: 'For, as it is said, we breed crows to pick our eyes out', which is a literal translation of a Spanish proverb, 'Cria cuervos y te sacarán los ojos'. Hudson knew this. There was an invisible wall between the languages. On hearing a gaucho tale about an encounter with an ass and a puma, that seemed so natural in Spanish, Hudson reckoned, 'it cannot be told here'. The title of his story 'El Ombú' was a sound poem in itself, an exotic Guaraní word for a sacred tree and simply untranslatable. The English equivalent might be the hawthorn or yew, England's oldest indigenous tree.

Hudson's listened to gaucho tales after a hard day's work on the pampas when there was nothing else to do. Then he recreated these oral tales as stories, poems and essays in English, but retained the sense of a voice speaking in an anecdotal style. You *hear* it. The poet

John Masefield caught this storytelling gaucho in Hudson. He evoked Hudson as tall, well-made and noble-looking, but remote and melancholic, as if he had just emerged from a thicket on to the pampas, where 'he had been free as the air'. Hudson loved telling of his gaucho days. Masefield claimed: 'I cannot convey the power of such story-telling: it was partly the beauty of the strange strong face that watched us as he told, partly the poignancy of what we felt to be his agony.'

Scattered throughout his writings are fragments of gaucho oral lore. In a letter he recalls the gaucho fear of the dark generating a monster with green teeth called the *'curapita'*, but this didn't become a story. In a letter to John Galsworthy of 19 May 1916 Hudson explained that his story, 'Pelino Viera's Confession', published in 1883 in the magazine *Cornhill* as his first story but rejected by Garnett for inclusion in *El Ombú*, was based on another gaucho story he had overheard on the pampas from an old gaucho. He reckoned it was his best story.

Shipboard friend Abel Pardo, from the *Ebro* trip in 1874, translated it for *La Nación*. Set in 1829 under tyrant Rosas and before Hudson's birth, it takes the form of a manuscript confession of Pelino Viera, a wife-murderer. The narrator's sole task was to translate it into English. Hudson saw himself in general as a translator of the vanishing gaucho way of life for the English. Pelino intuits his lovely wife is a witch. He consults an old witch-doctor, who prescribes some herbs. Superstition still rules in rural Argentina. Pelino then watches his wife anoint herself with a potion and become a bird. He follows her to an island on a lake called Trapalanda and in this dream journey wounds her. She dies back in his room. He does not feel responsible. Trapalanda is a mythic gaucho island where the dead end up. When Pelino knows he will be executed he voices Hudson's own exile:

Sometimes, lying awake at night, thinking of the great breezy plains, till I almost fancy I hear the cattle lowing far off, and the evening call of the partridge, the tears gush from my eyes. It would be sad to live far away from the sweet life I knew, to wander amongst strangers in distant lands, always haunted by the memory of that tragedy.

Hudson wrote his story in his tower eyrie in Paddington, haunted and moved to tears.

His poem 'Tecla and the Little Men', subtitled 'A Legend of La Plata', is also told by an old gaucho to ward off night fears. Horacio Velázquez, a biographer, argued that it was placed in Ensenada de Barragán, an old port from 1623 with a fort built in 1731 that later repulsed the English invaders, who then were forced to land in Quilmes. Swamps are also mentioned, as is Punta Lara, a poorish area with campsites overlooking the river sea, just beyond Ensenada. We walked around the ruined fort and sat out a tropical downpour. The fort and its five guns, a small affair, looked out over dense sub-tropical vegetation, the river now lost to view. The poem, set ninety years earlier, recalls a little girl and the evil dwarves who make her ill and carry her off to death. Hudson evokes gaucho territory: 'the plume-like grass / That waved so high', the ostrich, wild horse and deer. The topography corresponds to Hudson's family *rancho*, looking out onto the marsh and the site of an old *estancia* house where Hudson used to play. It belonged to the Lara family.

To help her ill daughter, Tecla, the mother tried out a diet of tiddlers from the stream, snipe eggs, curds with juice from thistle bloom, yellow roots, ostrich gizzards, but nothing worked. She called the rough monks from the nearby Dominican monastery (Santo Domingo again) and six arrived on horseback with holy water to bless the house. Then they feasted and Friar Blas sang with his guitar. But despite the hearty monks, Tecla was snatched by the evil imps. It's a fairy tale about the night forces that religion cannot assuage. The narrative poem ends on the 'deep silence of the night'. The poem is sung round a fire, where the immense night presses down eerily on the chatting gauchos.

Another anecdote recalled by Hudson refers to gaucho black humour. Once in 1915, he evoked again the small riverine port of Ensenada de Barragán, where, around 1864, he helped a friend load 6,000 sheep on to small sailing ships to ferry them across the sea-like river. It was excessively hot, and eight gauchos carried one sheep at a time on their backs on to deck. When they rested every hour, one of these men told such a funny story that 'amused me so much that I remember it to this day'. It was about the Age of Fools. A man goes off to see the world, returns rich with gold coins. His wife screams with delight and rushes to the wine cellar, knocks open a huge cask and drowns herself for pure joy; 'thus happily ended his adventures', the gaucho concluded.

As a boy Hudson had questioned his neighbour Barboza's refusal, even cowardice, to kill a man who had insulted him. He was told that Barboza was one of those fighters who fought to build his reputation and chose who to kill. He didn't kill because he lost his temper. Is this philosophy? Hudson learnt from a gaucho that nobody returned from death, that it was final. He also learnt about the value of religion from another gaucho neighbour with a mass of silver-white hair and recollected his actual words. 'You, Hudson, read books. We are Catholics and you are Protestants, now tell me what's the difference between us?' Hudson, feeling superior as a Protestant, answered. The gaucho kept smoking. 'Now I know,' he said. Then revealed his real thinking: 'These differences are nothing to me . . . because, as I know, all religions are false.' Hudson, aged fourteen, was shocked. The gaucho continued, 'your priests and our priests tell us we must believe. But there is no other world and we have no souls.' 'When the brain decays,' he went on, 'we forget everything and we die and everything dies with us.' Hudson listened in distress, haunted by this old man. As he wrote this piece he actually recalled the dialogue in Spanish with the gaucho. Hudson: 'Good reciters were common enough in my time'.

This is all so remote from us. Cowboys and Indians have become a Hollywood stereotype. But the meaning of the Indian frontier war is still disputed amongst Argentinian historians, where the Indians and their close cousins the gauchos have been decimated by the War of the Desert or incorporated into civil society. A frontier war had been going on for nearly 140 years. Hudson realized that his personal anecdotes about frontier life might not even be believed back home. The passage of time had made him a unique witness. We cannot even call them 'Indians' anymore, but indigenous peoples.

The city of Azul lies 299 kilometres south of Buenos Aires. It's the Cervantine capital of the Americas, as a Dr Ronco had collected countless editions and translations of *Don Quixote* or *Quijote*, displayed in his house, which is a museum. In one of the rooms I was surprised to see in a glass case Julian Barnes's donation of the first translation into English of the *Quijote* by Thomas Shelton, in memory of his late wife. Barnes had been a visitor. My wife and I caught the last day of the Cervantes festival and the opening of an exhibition of editions of Cervantes's *Novelas ejemplares*, accompanied by cartoons by Miguel Reps, a leftish cartoonist. There was a Museum of Modern Art hosting a performance artist who'd worked with Herzog's sound recorder. There were cafés with wifi and restaurants, including La Fonda, a favourite in a junk-shop atmosphere, with tapas that didn't stop dropping on our table. In Azul, the Indian past is remote.

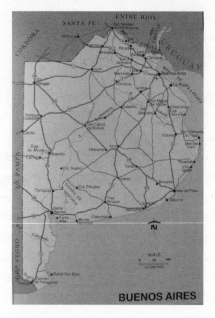

Map of Azul and the Provincia de Buenos Aires.

But for many years Azul was a frontier garrison fort, the furthest point south-west in Hudson's Argentina. It took nine hours by train. In an 1870 letter published in the *Zoological Society Proceedings* in London, Hudson mentions this frontier with its 'vast unexplored regions lying between the grassy pampas and the Andes'. He picked out the *butarda* (*Bernicla magellanica*), found in great numbers in this wild Azul region, feeding on the clover and tender grasses. His letter, read aloud by the society's director Dr Philip Sclater to a London audience, evoked these birds as 'shy and loquacious, and chatter much during the night in frost weather'. These beautiful black-and-white-striped geese, a pest for arable farmers, have a honking cry that's mesmerizing. Hudson admitted he had no idea where they bred. Their flesh was the most deliciously flavoured game bird that he had ever tasted.

Hudson played down, in his scientific way, the lure of 'unexplored country'. During his time Argentina was a country divided between the Republic, with its many people, and the nomadic, wild Indians. There was a frontier, but not as we know it today. Somewhere behind this long frontier cutting off half of today's Argentina lay the Indians and danger to life. On one side, the cities, ports, farms and progress, and on the other, a desert of long grass and flowering *cortaderas* and semi-naked, nomadic Indians. That was his Argentina. You could ride a horse over this frontier but you'd be aware of being spotted by Indian sentinels on horseback. It was frightening. As Lugones said, fear of the Indians was the 'devil' of the pampas. Just at this historic moment, landowners were greedy for this empty land and imagined filling it with cattle and sheep. Martín Fierro, the gaucho, noted how the rich yearned for this 'waste' land in *tierra adentro*. When Fierro returned from living with the Indians, he was made 'giddy' from riding over this monotonous grass. Until he saw an ombú and a shack, he wasn't aware that he had crossed over this frontier. In England Hudson continued to roam at will, over private land, as a poacher, in that spirit of danger and freedom that is at the source of his imagination, and of dissatisfaction with home.

In 1833 Rosas pushed the Indians out of their hunting grounds and established basic forts in Tandil and Azul. The meticulous note-taker William MacCann was in Asul, as he spelt it, in 1848. The frontier post was an assemblage of *ranchos*, with a guardhouse, a small church, a flour mill turned by mules, and roughly 1,500 inhabitants. The nearby Indians were not allowed to cross over the frontier. Just after Hudson left, a trench was dug called Alsina's ditch, an engineering folly supervised by Frenchman Alfred Ebélot in 1877 that never worked. It ran from Bahía Blanca to Italó, some 374 kilometres long. In 1879 General Roca, who would twice be president, set out with his Army from Azul, where the railway line ended. He was the first, in 1880, to establish the real frontiers of Argentina from the capital to the Andes and down south to Patagonia and Tierra del Fuego, liberating some 550,000 square kilometres.

The Argentina that Hudson knew from his herding and frontier days was a vast empty space, where no estates existed and land was not enclosed; open country, huge distances and semi-feral cattle. He evoked the pampas as *vacas y cielo* (cows and sky), similar to how José Hernández in 1872 has Martín Fierro suggest the treeless pampas: 'I could only see cattle and sky.' But defending land from marauding Indians, and then regaining it by force, happened *after* Hudson had left. In 1855 the ferocious Calfucurá (the name of an indigenous dynasty), with 5,000 horsemen with lances, sacked the fort in Azul, killing some 300 soldiers and settlers. After Calfucurá died in 1873, Namuncurá took over and in 1877, after Hudson left for London, 3,500 Indians attacked Azul again, killing 400 men, taking 500 white women captive and stealing some 300,000 animals. The Indians were real. Cunninghame Graham was in Azul in March 1876 and corroborated that 'this is a very curious little place. Formerly a tribe of tame Indians lived just outside the fort, but when the wild ones invaded the other day, the "Mansos" [tame ones] all retreated into the desert with them'.

Historian Moncaut unearthed fascinating documentation that on 8 June 1859 Hudson and his eldest brother Daniel were called up to serve in the Army, the Ejército de Guardias Nacionales, and had to report at San Vicente. In 1864–5 brother Daniel was soldier number 35 in the first squadron 29 Compañía de la Milicia Activa and transferred to Azul. But due to his poor health, Hudson was placed in the Milicia Pasiva as soldier number 76, also at Azul. In October 1864, according to research in the War Ministry by Luis Horacio

Vázquez, Hudson was called to carry out military manoeuvres in Rojas with Regiment 13 of the Guardia Nacional and became 'active' like his eldest brother. With Daniel, they were then sent back to Azul. By 1865 Hudson was back home with his ailing father.

His military service had lasted from six years. That is, Hudson had first-hand experience of the Indian way of life, briefly narrated as an episode in 'El Ombú'. He admitted he wasn't long enough among them to 'get through the outer crust'. But several months were spent on the frontier, that 'vast vacant territory'. He was not an officer, but a born horseman (foot soldiers were pointless on the pampas, given the immense distances).

The only presence of these Indians in Azul today is the ethnographic museum, with black-and-white photos of the sad, tamed Indians, who lived on in their slum tents, led by a dynasty of local rulers called Catriel. During Hudson's time, over 6,000 friendly Indians lived in a town of leather tents in Azul. Most of them didn't speak Spanish but Mapuche, so this linguistic frontier needed interpreters. MacCann devoted a chapter to these Azul Indians, their belief in the Sun God, their ear-rings and face paint, their horse skills and diet, some of which passed over to the frontier 'whites'. One item of Indian dress that became an Argentine icon was the poncho, woven by Indian women weavers. It is telling that they never laughed.

In 1872 a French doctor called Henri Armaignac visited Azul, keen to see the tame Indians. He quickly noticed that the Pampas Indians hardly knew Spanish. They used a Spanish interpreter whom they'd kidnapped for seven years. This Santiago Avendaño led Armaignac to the tall, obese and famous Indian leader, Catriel, sipping his *mate*. He'd been made a general after ceasing to plunder farms. He lived in a tent near Azul living off Argentine government supplies of tobacco, mare's meat and clothes.

Juan Manuel Blanes, *La Cautiva*, c1876.

By chance, a funeral was taking place. The passage of death for the Pampas Indians was a long horse journey, so twenty horses were slaughtered and placed in a circle around the burial mound. Death was seen as having been cursed so witches were rounded up and also killed. To accompany the dead man his youngest wife and one of their children were also suddenly killed in front of Armaignac, by a throw of a *bola*, a round stone with leather straps used to catch horses. This anecdote reveals how incompatible Hudson was with the frontier Indians.

In his story 'El Ombú', the *rancho* dweller called Valerio told Nicandro of an Indian invasion that besieged the garrison at Azul, swept into settled land and stole some 10,000 horses and cattle. These periodical Indian attacks involved semi-naked Pampas and Tehuelche Indians. On horseback with lances, greased in stinking horse and rhea fat, they rustled mares, which they ate, and kidnapped white women who became slave wives. Historian John Lynch called them 'fiercest of all the Indians of the plains' and 'irredeemably savage'. Borges's English grandmother once met an English woman who had been rescued but who shockingly chose to return to her life as a squaw. She saw her once more jumping from a horse to drink mare's blood as a man butchered an animal. Valerio was conscripted into the Army to chase the Indians. Like them, he survived on mare's flesh. The soldiers tracked the Indians to a valley near the sea. Valerio told it to Nicandro who told it to Hudson who tells us:

> Our troop of horses, urged on by our yells, were soon in the encampment, and the savages, rushing hither and thither, trying to save themselves, were shot and speared and cut down by swords. One desire was in all our hearts, one cry on all our lips – kill! Kill! Kill! Such a slaughter had not been known for a long time, and birds and foxes and armadillos must have grown fat on the flesh of the heathen we left for them. But we killed only the men, and few escaped; the women and children we made captive.

Hudson did his military service in the frontier Army. That was the crude reality of life for all in the countryside, but didn't affect those living in Buenos Aires. Once again, that city island avoided contact with its frontiers. The coexistence of nomadic Indians, who were terrifying horsemen, along a frontier about 200 miles from Buenos Aires, that ran through a series of forts like Azul and Tandil, wasn't resolved until 1880, six years after he arrived in England. This minor, gruesome slaughter summarized his Argentina. That's why gauchos and Indians acted as if 'human life is not worth very much'.

Hudson returned to frontier life with the untamed Indians in his posthumous book, *A Hind in Richmond Park*, his 'true testament' according to art critic Lombán. This frontier was defended by small mud-built forts at intervals of five to ten leagues. The Indians would burst through, kill the landowners and steal the animals. Hudson was aware that his English reader would find frontier life incredible, that the Indians didn't even have carbines or pistols, just very long bamboo cane lances. He admitted the Indians often won, despite their lack of technology, because they created terror in the soldiers' horses.

In 1922 Hudson told another story. A colonel was sent with 200 men to attack the Indians some sixty miles south of Azul. They rested at an abandoned ranch and herded their many horses into a corral (a rough enclosure for horses). The officer ordered his men inside too. The yelling Indians arrived and galloped round and round the ranch as the herded horses crushed the men inside. The Indians jumped off their horses and finished off the soldiers on the ground with their sharp lances. There was only one survivor, hidden under another dead body. This man told his story to Hudson. The cause of the stampede inside the corral was the stench of the Indians. They never washed or dusted themselves and were coated in rancid mare fat. Horses could smell them days before they arrived. Hudson added that this was a small incident, one of 10,000 little frontier tragedies and not of importance enough to find a place in history books.

In Carmen de Patagones in 1871, before the desert campaign or Indian killings, Hudson met a man called Sosa. El Carmen, as Hudson calls it, was founded as a fort in 1779 and remained a part of Buenos Aires province. Like all gauchos, Sosa had a 'preternatural keenness' and was the sole person, thousands of miles away in Patagonia, to feel the tremors of the terrible Mendoza earthquake of 20 March 1861. He claimed there were over 12,000 victims, but modern calculations decided 4,300 deaths. Sosa was also a horse thief.

Hudson was equally fascinated by Sosa's father, General Francisco Sosa, an Indian fighter under Rosas, who had expelled and killed

Indians along the southwest frontier, ensuring that Carmen de Patagones was Indian-free by 1836. This Francisco Sosa had chased Indian chiefs up to the island of Choele Choel in 1833, caught the notorious Chocorí, who, naked, slipped his guard, but left behind his fabulous cloak now in the La Plata museum. It's wrapped around a tall model of the Indian chief, with its red and cream colours, in a glass case. It's the first thing you see as you enter the large room. I'd expected it to be more luxurious, but for its kind and belonging to a nomadic rebel, it was fascinating. Chocorí, probably a Mapuche, was the leading cattle and horse rustler during Rosas's tyranny. He died in 1834. Sosa's notorious, cut-throat father died on his estate near Bahía Blanca in 1836, though Hudson claims Rosas had him poisoned, afraid of his prowess.

Another neighbour in Carmen de Patagones, called Ventura, a bore who drank and gambled, told Hudson an Indian story. He had just bumped into a childhood friend called Damian, after thirty years. He had been forced by the military commander to bring a troop of horses to an outpost. They found an empty corral on its own by the Río Negro bank and left the horses there when a group of Indian raiders appeared. They just had time to run into the river and swim. Patagonians are good swimmers, Hudson said, but one of them, Damian, tired and had to surrender to the waiting Indians on horseback.

Usually they kill male captives, but he eloquently persuaded them he hated the Christians and was an Indian at heart. They accepted him in their tribe and he lived in the apple groves of the Andes foothills. From these apple trees left behind by missionaries in the eighteenth century, Indians made their alcoholic *chiche* (Hudson calls it *chi-chi*) and had their proverbial week-long drunken orgies. Damian was given a wife and had children. But he yearned for home.

After thirty years of being a captive and the butt of Indian mockery as the white fool, but thoroughly 'Indianized', he escaped back to El Carmen. '"And there he is," concluded Ventura, when he had finished his story, with undisguised contempt for Damian, "an Indian and nothing less!"' Yet Ventura sat cross-legged on the floor, brown as leather and called himself a white man.

Hudson transcribed this tale in 1893 whilst living in Paddington, imagining 'the mysterious haze of the desert'. He was so touched by what he'd heard of this doubly homeless man, who had left his Indian squaw wife behind waiting for him and was an alien in his hometown, that he ended with 'Poor Damian and poor wife!'

The story 'Marta Riquelme', also published in *El Ombú*, is set in the northern Andean town of Jujuy and the nearby village of Yala. There's a consensus that Hudson had travelled there during his lost Argentine years, but no documents confirm this. Perhaps it was his elder brother Edwin, who lived there, who told him this tale. The background geography and climate to this story seem genuine. At least, when we visited Yala's small lake in 1973 to stay in an empty, crumbling hotel built for workers by the Peronist trade unions in beauty spots all over Argentina, I was aware how well he'd caught the bleakness of this high-altitude outpost.

The story explores the Indian mind. Here the Indians speak Quechua, live on *patay*, a sweet paste from the algarrobo or carob tree, and hang on to their ancient Inca and pre-Inca gods and demons. The unhappy Marta at the end is turned into a foul, shrieking bird-woman, Kakué, a corruption of the Indian word for the town Jujuy. Hudson had anticipated his Rima story of *Green Mansions*, another sad bird-woman. Marta's gambling white husband is conscripted to fight the Indians on the frontier. She sets off from Yala to join him with their child. After her mule caravan is ambushed by Indians and her child snatched from her, she becomes yet another captive. She escapes and is caught again by her Indian husband, who ties her to a tree, scourges her for a year, then frees her with a log tied to an ankle, while she gives birth to three more children. In case this cruelty is seen as far-fetched, Lugones describes a common method of stripping the sole off the feet of captives to stop them walking away. Captives were war booty. Marta escapes again, with her youngest, but as she struggles with the Andes and deserts, this child is chucked into a stream by the Indian helping her escape. She reaches Yala after six years with the Indians, but has so changed, is so burned by the sun, with wrinkles of grief and so skinny that her husband refuses to recognize her. The rest of the tale deals with the priest fighting the Indian demons that remain there despite centuries of Christianity.

One last story from *El Ombú* is the Spanish-titled 'Niño Diablo' (Devil Child), a frontier tale from what Hudson called 'the heart of the vast mysterious wilderness'. The anecdote is slight – a young boy kidnapped by Indians and all his family killed. He grows up familiar with Indian customs, language and horse skills and manages to escape. He knows how to steal horses and how to stop dogs barking when approaching a *rancho*. An old rancher and his attractive twin daughters are sitting by an open cooking fire when this '*niño*' sneaks in. Another stranger drops in looking for the *niño* and begs him to rescue his wife, kidnapped by the Indians. The *niño* agrees, outfoxes an aggressive neighbour, steals his horses and rescues the woman from the tents. Behind the story lies Hudson's wild life in that distant land – Tandil and the Arroyo Langeuyú near Rauch are mentioned – close to the Indian frontier, with constant talk of invading savages with lances.

Robert Cunninghame Graham, who boasted of his outdoor deeds, dedicated his facile biography *Hernando de Soto*, 1912, to Hudson: 'He knows the disappearing Indians, and can feel with them', for by then they had been decimated. The Indians were so hostile that nobody during Hudson's boyhood bothered about their way of life or studied them. Hudson once found a small piece of black pottery when a ditch was dug near Los Veinte-cinco Ombúes, made of baked earth with fingernail incisions. He preserved this talisman as a reminder that the 'vacant land' had been peopled with long dead and forgotten men. Hudson had smelled them, admired their horse skills and spent hours as a naturalist with the Keens, friends of his parents, at their ranch, Pedernales, one of the oldest in the Azul region, and along the Arroyo de Los Huesos and the river Callvú.

Hudson's view of the Pampas Indians was an insider, experiential one. He doesn't ever refer to Sir Francis Head's 1826 account. But he knew his Darwin, and Darwin often did. Head had found these Indians 'really wild and beautiful' and their air of 'unrestrained free-dom' exhilarating. He came across them as he rode the pampas and asked his readers 'and what does the civilized world know of them'. Hudson could fill him in.

When Col. Levalle, who led General Roca's forces in 1879 in the *Guerra del Desierto*, wrote to his commander that the chief Namuncurá had fled to the Andes and that 'no Indian tents remained', he announced the final cleansing of the pampas. Hudson twice told the artist William Rothenstein, while sitting for a sketch, that he had Indian blood: 'I am a red man, or at all events a wild man of the woods'. I doubt that his DNA would reveal American Indian blood (remotely possible through his mother's line), so this confes-sion is more to do with his identifying with those most distant to Rothenstein in his Hampstead studio.

Did Hudson ever kill an Indian or a gaucho? Some biographers think so on the evidence of the character of Richard Lamb in *The Purple Land that England Lost*. Lamb ended up in a *pulpería*, where one of the long-haired, murderous-looking men recognized him as a rebel fighting against the government, and held him up until the police arrived. Lamb then tricked this man and shot through his poncho and killed him. He had awoken the instinct of self-preservation in 'all its old ferocity', with such implacable aggression that he even kicked the corpse as he leapt onto his horse. As he galloped away he experienced such 'joy' he could have sung aloud. My guess is this was novelistic convention, rather than Hudson's secret confession from his gaucho days.

Hudson didn't write at length specifically about the *indio* or *gaucho* – his thoughts are scattered throughout his anecdotal writings – but about wilderness life in general. Nonetheless, as I wandered through Argentina over the decades, criss-crossing Hudson's own trails, his stories helped me sense the distant, wild and increasingly forgotten life in the pampas. The blending of *indio* into *gaucho* and then *peón* in Argentinian country life is total, as are many of their shared horse habits, as well as their reluctance to speak directly to you. Always saying what you want to hear. Indigenous women were sold as servants and whores to the male immigrant masses and thus perpetuated their genes. I find that Argentinian society is axed on a suppressed racist basis. At the top of the social pyramid are the fair-haired (and dyed blondes) and the lower you go down the darker the skin. There's no official racism, but scratch a *porteño* and he or she will rant on against a Bolivian or a Paraguayan. Racial differences are more accepted in the rest of Latin America, but still there's discrimination. Everywhere you go in Argentina you see signs of straight black hair or beardlessness, just as in southern Spain and southern Italy the ancient Muslim past breaks through. But in Hudson's time, before mass immigration, racism was more relaxed. I come from a white Mauritian family petrified of two racist possibilities. One was becoming a 'poor white', so my ancestors worked hard to avoid this pitfall. The second was having 'coloured' blood, which you just hid. I laugh at these ancestral fears today and

even find the reverse is desirable. But my father stopped speaking with his first cousin, a judge, who had written a booklet saying that our family had 'coloured' blood. A great grandmother did look vaguely 'Eastern' or Lebanese, and the children of this first cousin could pass off as Indians from India. This racist tinge drove my father mad. He even married a Norwegian (as did his cousin). I remember an anecdote my mother told me about that when she was engaged in Nice in the 1930s, the English vicar had told her stepmother that the 'Wilsons' had coloured blood. This possible accusation of having Indian blood was the skeleton in my father's family cupboard. It would come as no shock to Hudson that 54% of Argentinian DNA is indigenous. For him it was a natural outcome and he saw it in his neighbours in the pampas. He was so relaxed about this that he even identified with it once in England, to shock his English readers.

The autocratic and cruel Juan Manuel de Rosas (1793–1877) featured in Hudson's family life. When Hudson travelled to England in 1874, Rosas was still alive and living in exile near Southampton. As he drove around in a trap, Hudson came across Rosas's 'modest thatched cottage', with its orchard and green fields, where he was 'chiefly known for selling fresh milk at twopence per quart'. He didn't stop and call on him, yet Rosas's portrait had hung in the Hudson family hut.

From his birth in 1841 to when the tyrannical rule ended in 1852, Hudson lived under Rosas's shadow. When about six, Hudson was taken by his mother to Buenos Aires. His account in *Far Away and Long Ago* evokes his stay near the riverfront, with boats anchored far out as the water was shallow and there was no harbour. He especially remembered his glimpse of Don Eusebio, Rosas's dwarf court jester. He was dressed as a general in scarlet, with a three-cornered hat and large scarlet-plumed aigrette. He marched past with twelve soldiers as bodyguards. Nobody could laugh or mock as they would be killed.

John Masefield included this fool in his narrative poem *Rosas*, 1918, 'dressed like [a] British general' who kept 'some twelve, his chosen murder-ban'. Hudson had told his Rosas story to Masefield, but complained in a letter to Garnett that the poet's South America was too imagined, too far away and false. Masefield held a 'deep admiration' for Hudson, like all the young writers of his time. Incidentally, whenever they met, Masefield sensed that he had been visited 'by the unearthly'.

Hudson never met Rosas, unlike Charles Darwin, who found him 'extraordinary', a 'perfect' horseman. Hudson's attitude to him was ambivalent. Rosas was the Nero of South America, but also original-minded, the 'greatest' who had ever climbed to power. In 1919 in a letter to Cunninghame Graham he praised him as 'that tremendous man'. Nothing about his secret police, the dreaded, throat-cutting *Mazorca*, or having to always wear red ribbons in public, or the intellectual opposition driven into long exile in Montevideo. Being loyal to Rosas went even deeper. Men had to

Emeric Essex Vidal, 1819, watercolour of Fort and washerwomen.

display long sideburns and moustaches. The nightwatchman called out the hours, adding 'Death to the savage Unitarians' (Rosas's enemies). Every official document had to state the same message. There were spies everywhere, especially the black servants and washerwomen, but Hudson offers nothing on these codes of practice and no sense of what exactly this Nero had done. We share a boy's delight in a spectacle, not an adult's political analysis.

Hudson's father was proud of the family's portrait, in a heavy gold frame, of this dictator with light reddish hair and piercing blue eyes, sometimes known as the Englishman or as the Farmer. The Hudson family, like most English merchants and farmers, were *rosista* – followers of Rosas. English businessman Charles Derbyshire claimed that Buenos Aires had never been safer than under Rosas. The issue of safety in the streets is today paramount. I recall that Spaniards said the same about life after Franco. The streets had never been safer. But not only Rosas's portrait hung on the wall.

Hanging next to Rosas in the drawing room was his wife Doña Encarnación Ezcurra, a handsome, proud-looking woman with a vast amount of black hair. When she died in 1838, Rosas imposed two-year compulsory mourning; everybody had to wear black, but were the Hudsons excluded from wearing black? I don't think so and it's not mentioned. Doña Encarnación ran Rosas's spy rings and made lists of enemies to kill. She was whimsical, but visited and organized the poor. In a letter of 1916, Hudson recalls a strange detail that he had heard about Rosas forbidding a priest to give Doña Encarnación the last rites when she was maddened by remorse and in terror of God's wrath.

Next to her on the wall in Hudson's home was a portrait of Rosas's ultimate traitor, General José Urquiza (1801–70), a 'ferocious cut-throat' who was murdered at his extraordinary palace San José in Entre Ríos while Hudson was still in Argentina. Hudson saw no good in this man. He knew all about Urquiza as news spread by voice beyond the cities. But did he care? Certainly when he wrote his memoirs he had forgotten Urquiza's role in Argentinian politics. Or perhaps being a foreigner meant he wasn't involved in local politics. Was Urquiza more than a ferocious cut-throat? He was; and Hudson got Urquiza wrong, or saw only one side of the man. He was a strange, tough man, who added to the family's wealth by farming and owning a huge 'salting plant', before going into the wars and politics. He ran the state of Entre Ríos like a feudal one, ending gaucho anarchy. He was a Don Juan, with countless illegitimate children (but all acknowledged in his will). He was methodical and austere. All his foreign visitors noted that he didn't smoke, drink alcohol and most oddly, didn't touch *mate*. Politically, he was a fierce defender of a federal state, against the capital Buenos Aires, and thus loyal to Rosas. As a general under Rosas, he defeated the governor of Corrientes, followed by several days of throat slitting (*degüellos*) described by historian Carlos Ibarguren as 'unrivalled in gaucho barbarity'. In another victory at India Muerta in Rocha, Uruguay, he ordered that the five hundred prisoners have their throats slit and posted a note that said 'whoever buries one of these will have his throat slit', in case any showed pity. So Hudson was right.

But once in power, he banned political executions, forbade land confiscations and fomented agriculture by stopping the import of wheat. He was an ardent educationalist, founding a famous college in the town of Concepción on the Uruguay river. He brought in an eminent French director, Alberto Larroque. A contemporary (and ancestor of my wife's) Vicente Quesada was impressed by his prodigious memory, knowing the names and families of most of his Entre Ríos gauchos. Mariquita Sánchez sized Urquiza up well: 'I find this man all the greater as, so they say, he never went to school, never read a book. He is pure instinct and nature'. She again confirmed that gauchos had incredible memories because not deformed by education.

After defeating Rosas in the battle of Caseros in 1852, he didn't seek revenge and later let Rosas sell his estancia San Martín in Cañuelas in order to buy a small farm near Southampton, England. No winner, no losers was Urquiza's victory motto. He became president in 1854 and moved the capital to the small riverine city of Paraná. He finally defeated General Mitre at the battle of Cepeda in 1859, when Buenos Aires agreed to rejoin the Confederation. That was his good side.

He had already survived two assassination attempts, but there was a third, orchestrated by federalist Ricardo López Jordán, who later alleged he had never meant to have him murdered. But he was a traitor to the federal cause. On the 9th April 1870 some fifty armed men burst into his palace, shouting 'Death to Urquiza, traitor'. He was shot by a pistol in the right cheek and then stabbed five times to death in his bedroom by a former protegé. Robert Cunninghame Graham was there in the Palacio San José soon afterwards and told his version (at odds with the official trial): that Urquiza was drinking *mate* (but he didn't touch it) when a gang of men burst in. One of them, with only one eye, drew a dagger and shouted death to all tyrants. While his daughter had found a pistol and shot one of the attackers, another, the same one-eyed thug, stabbed Urquiza in the back. He was then hacked to pieces on the sofa, blood all over the marble floor. Two of his sons were also assassinated.

I saw Urquiza's bloodstained fingerprints on the wall in the Palacio San José's ballroom where his assassins stabbed him these five times to ensure he died. Or at least they're painted in. I walked about this extraordinary palace rising up in the middle of nowhere. It has two towers, or miradores, one with a working clock and the other with a clock stuck on the time of his murder. Someone stopped that clock, that is. It also has a French park replete with statues, palm trees, Brazilian araucaria and two lovely flag-stoned patios. He planted manioc, maní, apple and pear trees as well as countless flowers, with a Frenchman as his head gardener. Urquiza had imported marble and artisans from France, Italy and Uruguay to create his bizarre dream palace. The ornate chapel has murals by the Uruguayan painter Juan Manuel Blanes (1830-1901) around its blue-tiled dome. In 1856 Urquiza commissioned him to paint eight huge oils depicting his battles like Pago Largo, Sauce Grande, India Muerta etc. Blanes avoided the blood, gore and suffering. These were heroic, cleaned-up accounts, not documents. In 1858, Blanes painted the murals in the Palacio San José's capilla. He lived on the spot for nine months. Sarmiento much admired this patriotic realism. When Sarmiento stayed he had to listen to one of the twenty-three of Urquiza's recognised children play the violin in the mirror-ceilinged drawing room. Urquiza loved watching himself dancing.

The pergola of ornate metal, the first house with hot running water in Argentina, the complicated drainage systems, mirror-heated birdcages, all showed a fertile mind intent on controlling every detail. When Urquiza finally retired to his palace, the poet Gutiérrez said to him that paradise would be a week at San José chatting with the General under the trees and eating peaches. He had realised how exceptional it was to have that happy kind of garden and palace in a war-torn country and times. Urquiza never travelled to Europe, but brought his idea of Europe as architectural fashion to Entre Rios. I found this palace surreal.

There were two further portraits; one of General Oribe, who besieged Montevideo, where Rosas's enemies had fled; and the other, unnamed, but in fact of Rosas's Minister of War, Angel Pacheco, a rich, neighbouring landowner. What a picture gallery of thugs in power hung in the Hudson family home.

In 1919 Hudson corresponded with a son of Manuelita, Rosas's daughter. She had sailed out to a Southampton exile with her father in 1852, and married Máximo Terrero in England. Juan Terrero, their son, was 'brought up as an English gentleman' and had read about Rosas's dwarf Eusebio in a magazine, before he appeared in book form in *Far Away and Long Ago*. Hudson recalled riding over the vast farm lands the Terreros owned in Argentina, and wondered what this anglicized gentleman thought of Rosas, his sometimes cruel and murderous grandfather.

Hudson recalled hearing the din of the battle of Monte Caseros in 1852 when Urquiza defeated Rosas. The musket shooting was like distant thunder. Rosas's defeated Army straggled past the rented home, Las Acacias, in their scarlet uniforms. These soldiers had captured an unnamed officer, who sat quietly on his horse. He then tried to escape, but could find no protection in Chascomús and was recaptured. Later Hudson was led by a native boy to see a great stain of blood on the short dry grass where the young officer had had his throat cut, the sinister gaucho execution systematically carried out in the cruel and violent civil wars. It was an Indian method.

The gaucho *facón* knife was used to kill enemies, to slaughter sheep and cattle, and to eat roasted meat. This wide knife is a potent symbol of manliness. During the civil wars of Hudson's youth there was no taking prisoners to non-existing prisons. The *facón* still defines the *peones* who work on the farms. I sat once at a long trestle table with about sixty men and my mother-in-law for a lunch in Rauch after a cattle auction. Nearly everyone took out the *facón* from their belts to eat the meat, and left the knives and forks untouched. Readiness to use this knife led to constant outbursts of violence. When I first visited the main family estancia La Isabelita, before the break up of the large farm, I was sitting in a doze after another of the interminable meals, a cruise ship lifestyle with five meals a day, when the uncle who ran the farm rushed through the drawing room being chased by the cook's raging son, with a *facón* in his hand, and himself being chased by the farm manager. They ran past me into the night, where he was disarmed and sacked on the spot.

In Hudson's Argentina the *facón* suggested more than a usual hunter's thrill of the kill: 'And in those dark times . . . the people of the plains had developed an amazing ferocity, they loved to kill a man not with a bullet but in a manner to make them know and feel that they were really and truly killing'. Later, when Hudson worked the land, he heard more about this blood-lust. He overheard talk that turned with 'surprising frequency' to how soldiers cut throats so as not to waste bullets. They chose young throats and performed their murder 'in a leisurely, loving way'.

In his *Voyage of a Naturalist* Darwin had praised the gaucho for slitting throats in a gentlemanly way. Hudson corrected the grand

Daguerreotype of Camila O'Gorman.

scientist: 'even as a small boy I knew better – that he did his business rather like a hellish creature revelling in his cruelty'. Hudson recalled the actual words of these ex-soldiers boasting of their deeds as they murmured to their victim: 'Think of the sight of warm red blood gushing from the white column!' Fifty years later in 1918, that memory still caused a spasm of nausea in him.

Despite Rosas's cruelties, Hudson admitted he retained his father's faith in the leader as 'restorer of peace'. But he recalled one horrific case of Rosas's 'inexplicable' crimes, the public execution of a young lady and her priest lover. He doesn't mention their names, but it was the main scandal of Rosas's years in power. He was six years old when it happened. The beautiful Camila O'Gorman ran off with her lover priest, Uladislao Gutiérrez, in December 1847. They had fallen in love while he was a priest in the Socorro, a simple eighteenth-century church in Buenos Aires on Suipacha and Juncal. The couple settled incognito in Goya, on the river Paraná, and ran

Domingo F. Sarmiento by Eugenia Belin Sarmiento,
Museo Histórico Sarmiento.

a school for five months until betrayed by a passing priest. It was the
first school in that village.

After she had been brought back to Santos Lugares (Rosas's military HQ, barracks and prison), Rosas's daughter Manuelita pleaded
for her, and arranged for books and a piano to be installed in her cell,
but Rosas was in a hurry. Within two days, on 18 August 1848, he had
the lovers shot. Camila was eight months pregnant. It was a triple
crime in that the unborn baby was also baptized.

Camila had such heavy chains on her feet she had to be carried
to the execution wall and placed on the seat. The first volley of bullets missed and one soldier fainted. The second volley saw one bullet
slice her arm and she screamed, hiding behind her long hair. The
third volley riddled her with bullets. The officer in charge went mad.

An 1848 pamphlet justified her execution because she had trans-
gressed the social codes. Her own father, in a letter of December
1847 to Rosas, asked for the firmest punishment.

Sir Horace Rumbold, British Chargé d'Affaires, called it 'the blackest of all crimes greatly hastening Rosas's downfall'. Domingo Sarmiento, later president, commented that all Buenos Aires 'froze in horror' at the shooting. He noted that both lovers were buried in the same coffin. Camila O'Gorman became a symbol of the tyrant Rosas's inhumanity. He had commanded her execution because he could not tolerate 'licentiousness and disorder'.

John Masefield's narrative poem *Rosas* ends with this execution of Camilla (spelt with a double 'L') and the Chaplain Laurence (he also got his name wrong), so that their death becomes the battle-cry against Rosas, who 'died in England'. Hudson had told Masefield about Camila and her fate. But, despite this tragedy, in a letter to Cunninghame Graham, Hudson joked: 'I join with you in a Viva Rosas!'

In his appendix to the *El Ombú* collection, Hudson explained Rosas's appeal. For despite his immense wealth (by 1852 he had accumulated 816,000 acres) and belonging to the distinguished Rozas family (Rosas changed the spelling), he was 'by predilection a gaucho' and adopted the 'semi-barbarous life of the plains' that was engrained in Hudson too. Rosas was a daredevil, leaping bareback onto wild stallions. Hudson added: 'He had all the gaucho's native ferocity, his fierce hates and prejudices'. His intimate knowledge of the people allowed him to rule over them.

Chapter 4

Illness and the Making of an Autodidact

So much for the intrusions of Argentine border history on W. H. Hudson's personality. The shock of ill-health also leads to the inner man, for he suffered poor health all his life. Throughout his correspondence he moans about feeling under the weather, but he lasted eighty-one years. He was no *malade imaginaire*, though he often used illness as an excuse for not socializing.

He told John Galsworthy that the interesting part of his life ended when he ceased being a boy, but he didn't mean the onset of puberty, rather that illness forced him to be an invalid, stay indoors and take up reading. His story about his illness in *Far Away and Long Ago* is titled 'Boyhood's End'. His first fifteen years had been instinctive, outdoors and packed with first-time experiences. When he reached fifteen, Hudson spent four weeks in Buenos Aires, staying with the Methodist pastor on Bolívar 314. He went fishing, watched the laundry being done by black women on the shore, visited the

Carlos Pellegrini, *Carro aguatero* 1830.

great South Market with countless caged birds, but was not aware that he was catching a near-fatal disease.

He rode home the twenty-seven miles in a strange state of lassitude, then one night collapsed unconscious. There were no doctors, so his mother attended him. He had caught typhus, a bacterial disease brought on by lice or fleas with a 60 per cent death rate.

Hudson deemed Buenos Aires the most 'pestilential' city on the globe. Drinking water came from wells or from men on horseback, who carried water to the houses, or from *'aljibes'* (he misspelt that as *'algibe'*), rainwater tanks under the houses. Hudson recalled sixty-two years later as he was writing this passage that 'you always had one or two to half a dozen scarlet wrigglers, the larvae of mosquitoes, in a tumblerful and you drank the water, quite calmly, wrigglers and all'. Since that disease in 1854, as Hudson himself noted, Englishman John Bateman had solved the sewer and drinking water problems to finally dispel Buenos Aires's cloacal stench.

Nearly dying aged fifteen afflicted Hudson deeply, but all around him he witnessed death and killing. The gauchos knew that life on the pampas was not worth very much. 'What does it matter?' said one to him. But Hudson's own mortality shook him. Then, at the age of sixteen, as explained in the penultimate chapter of *Far Away and Long Ago* titled 'A Darkened Life', he was struck by a second serious illness.

The family was back at the original *rancho* in Quilmes and he carried out the tasks that a small farm demanded. He drove a herd of wild cattle in rain and wind, got soaked and exhausted and caught rheumatic fever.

This fever led to a lifetime of palpitations. After seeing all the doctors available, he was told that he had a permanently bad heart and could drop dead at any moment. A police doctor (and writer) at Quilmes, José Antonio Wilde, confirmed in 1870 that 'D. Enrique Hudson suffers from an organic vice, undergoing frequent rheumatic attacks'. Illness might have kept Hudson from frontline military ser-vice and war with Paraguay, but it didn't prevent him from sending letters and specimens to London's Zoological Society.

This disease prompted lawless thoughts about God, the afterlife and personal annihilation, as he put it. The direst consequence of his adolescent illnesses was grasping that every man 'must die alone', that his most intimate thoughts and his deepest emotions reflected his aloneness. The everlasting 'apprehension of loneliness' never loosened its grip on him. It's the key to the man that, in his words, 'I never had nor desired a confidant.' He had learned to face ill-health alone: nobody else could suffer the loneliness that constant rheumatic fever induced.

The second factor that conditioned Hudson's uniqueness was that he never went to school, let alone university. He was so relieved by this that he later called schools mind-killers, where too much reading degraded the student. Reading books was always associated with a 'pale phantasmagoric world, peopled with bloodless men and women that chatter meaningless things. The feeling of unreality affects us at all times'. All his life he loved reading, but it was something done in the evenings or during winter when he couldn't be out of doors. But it was a waste of time. In Cornwall, he would rather watch a sparrow than read about one. To read, he wrote, 'is to imbibe false ideas' and he preferred 'observing for yourself'. He detested the false authority of book learning.

There were different tutors at home at Las Acacias who were employed haphazardly when they happened to pass by, but in essence his mother and his elder brother taught him. Hudson has been accused of not being interested in people, but his pen portraits of his tutors in *Far Away and Long Ago* are vivid. One was Mr Trigg, fat and jovial, but who had to leave the 'dry' family every weekend to booze. He was begged not to use physical violence and complied until he lost his temper and set about the children with his horse whip and was sacked.

Mr Trigg brilliantly impersonated the characters in the Dickens novels he read aloud in the evenings. He was interested, Hudson wrote, in culture, at odds with the English philistines around and taught in order not to get a proper job. He was an actor who once dressed up as an old Scottish lady and fooled everybody. Another tutor was an easy-going Irish priest, Father O'Keefe, who proved to know no maths – it was he who was denounced by brother Edwin as a fraud.

The Hudson children, isolated as Americans, turned inwards and tried to run a family newspaper called 'The Tin Box', like the Brontës in the Rectory in Haworth, Yorkshire with their imaginary kingdoms. But it didn't last as the boys were too anarchic to collaborate. But they did sit together and have the Bible read aloud by mother everyday. It was the mainstay of their education. They all knew it by heart. As with many others, the rhythms of biblical prose entered Hudson's psyche. But, insatiably curious about the natural world, he read all the old natural history books he could get, until he was given Darwin's *On the Origin of Species*. It doomed his mother's small library of religious and scientifically dated books. Later, libraries always 'oppressed' him.

Self-education separated Hudson from his contemporaries in England. In a letter to the Ranee of Sarawak of 5 January 1914 – he was seventy-three years old – he mentioned that he was writing a testimonial for his younger friend, the poet Edward Thomas. 'How funny!' he wrote, 'A wild man of the woods giving a testimonial to an Oxford graduate', but it was written without the grudge of Thomas Hardy's *Jude the Obscure*. He contrasted himself with Thomas in a piece he left half finished when he died in 1922 as 'unschooled and unclassed, born and bred in a semi-barbarous district among the horsemen of the pampas'. It's a boast, for he was free from educational prejudices as he grew up among illiterate gauchos, where book-learning was an oddity and there were no social classes. In 1869, 79 per cent of Argentines were still illiterate.

In a letter to Linda Gardiner, Hudson referred to his namesake W. H. Hudson, a professor of Literature at London University, and asserted: 'My learning is on a par with that of a street arab who has succeeded in dodging the school inspector.' However, the family's educational enterprise succeeded because his younger brother Albert Merriam became a teacher in the Colegio Nacional of Buenos Aires and older sister Caroline, a nanny, in a census called herself an 'educationalist'.

Hudson associated books with being trapped indoors and illness, and later with living in London and economic necessity. While he rode the pampas or later rambled around southern England he took notes, but didn't write. As most of his work came from direct experience, note-taking was, after reading, what made a writer of him. He warned his friend Morley Roberts to make notes as 'memory is no good'. Without notes, he wrote, 'the keenest observation and the most faithful memory are not sufficient for the field naturalist'. And the same applied to the writer. At one moment in *The Naturalist in La Plata*, to illustrate spider gossamer, Hudson actually copied out his exact notes (in surprisingly clear syntax). But few of the thousands of notes survived his back-garden bonfire. When Cunninghame Graham took up writing, he lamented he hadn't made notes and relied on vague memories (that's the difference between their writing). Notes and memory mutually enhanced each other. Jotting words down as you ramble is as close to the initial experience as it's possible to get and keeps freshness alive.

The most vivid books that Hudson read instead of a formal education were about an imagined England, a dreamland, ages before he stepped on its shore. I was too young to be seduced by this ideal land from our colony, Mauritius. Hudson travelled to England aged thirty-two and I was seven. I see now that my parents' dreams were created by distance and ignorance. That their England was stuck in a pre-war structure. My mother idealised social life through reading English novels, without taking the rigid class structure into account, and suffering it. My father struggled with the climate and lack of light in dismal paintings of urban roof-tops and chimney stacks, before exploding into the colours of abstract expressionist art. In one sense, the explosion of colour from the 1950s in his art is a recuperation of the tropics. Then he turned inwards into scientific books and debated life from there. He could have been anywhere. He had eliminated his roots, as Hudson did.

From Argentina, Hudson wasn't clear whether prose or poetry enabled him to see reality more sharply. He remembered buying his first poetry book in Buenos Aires in the early 1860s, in a second-hand bookshop in the city's south run by an 'old snuffy spectacled German'. Amongst the books on the shelves and heaped on the brick floor was James Thomson's *The Seasons* (1726–30). Hudson had no idea who the poet was or that Wordsworth valued him highly. He never forgot the thrill of his first bought book.

In the same bookshop he found Robert Bloomfield's *The Farmer's Boy*, 1800, by a humble farm boy – a peasant-poet – and based on the farmer's tough seasonal life. Later, Hudson made a pilgrimage to Troston Park in Suffolk, where Bloomfield had lived. His essay on Bloomfield is a defence of those lesser poets who created his 'spiritual country'. His England, the 'land of my desire', was rural and timeless because he had read about it from Argentina. He identified with the boy of the poem as both had been shepherds: 'How often . . . I used to ride out to where the flock of one to two thousand sheep were scattered on the plain, to sit on my pony and watch the glad romps of the little lambs . . .'

Patrick Dudgeon, whom I met at his teaching institute in Buenos Aires and who was friend to Lawrence Durrell and Gordon Mayer, also made a pilgrimage to Troston, in Hudson's footsteps. He reckoned that Bloomfield was Hudson's guidebook to his imagined England. He'd brought it with him on the *Ebro*. When, after his death, Hudson's literary heirs sold his library, there was his 1820 copy of Bloomfield.

Hudson's 1896 essay 'Selborne', placed as the concluding chapter in *Birds and Man*, 1915, is his account of a fifteen-mile pilgrimage in the rain. Selborne, a little old-world village at the foot of a hanger and beech wood, has been preserved as a shrine to a bucolic England that was, and it hasn't been allowed to change that much even today. England does preserve its past in an artificial way by not allowing architectural progress. Charities like the National Trust try to keep this past alive, but the loss of the actual is there to be seen. My visits to Selborne tap into this ambiguity. The woods above are similar and are not. The village has kept a veneer of Gilbert White's days, some pretty houses, the church, the cemetery, but village life has changed. Not knowing how to react to this 'ideal' England, I simply uprooted a tiny holly from the hanger above the village. It has been growing in my London backyard for over thirty years. Every time I touch this red-seeded holly, I'm reminded of White and Hudson. Of course, the point of the essay is Gilbert White, the yew tree and his book *The Natural History of Selborne*: clear, delightful letters to friends from this naturalist curate who grew up in the village and whose modest tomb stone – still visible with its 'G. W. 1793' – summarizes his life. His calm letters appeared in 1789 as the French Revolution burst out. Selborne in White's time was isolated from the rest of England, already an Arcadian eddy.

White jotted down all that caught his eye with special concern for the swallow family, indeed all birds, in what he called his plodding manner, with few books and nobody to share his discoveries except through correspondence. He initiated Hudson into this direct, emotional grasping of the minutiae of natural history, without experiment or abstract theories. What separated Hudson from White was the guiding principle that God had beautifully designed his creation. One word summarizes this difference: Darwin.

Hudson confessed that White was continually on his mind. It was while pondering White's incessant curiosity that he came up with the core of his own ecological thinking: 'All the same, facts in themselves are nothing to us: they are important only in their relations to other facts and things – to all things, and the essence of things, material and spiritual.' Hudson was seeking 'something beyond and above knowledge'. For through White's book, 'chock-full of facts', shone his anxious-free personality. But the essential point is that he read White's notes on his parish in his teens before reaching England and it anticipated his England of the mind.

In 1929 poet Edmund Blunden, in his *Nature in English Literature*, established White as Selborne's Virgil, leading his readers through a contemporary hell and epitomizing an ethics – observing local worlds within worlds without egotism, but with an alert curiosity, a spirit of wonder in sober prose music. The pre-Darwinian outdoor naturalist was a saviour in a moment of threatened national character. White was the complete, rural Englishman. So Hudson fused his admiration with Blunden's nostalgia as the English countryside was changing.

But White did not deal with all Hudson's questions. Hudson remembered the day in 1856 (he was fifteen) when an English merchant friend from Buenos Aires brought him White's *Selborne* as a present. He read it many times but White's book 'did not reveal to me the secret of my own feeling for Nature'. White was too matter-of-fact, while Hudson was experiencing unaccountable mystical raptures as he asked himself, 'What does it all mean?' He turned from White's *Selborne* to Brown's *Philosophy* (in his home library) to define his personal brand of mysticism as being 'lifted out of himself at moments'. I think he meant Sir Thomas Browne's marvellous essay *Religio Medici*, 1642, as his memory often confused titles. He established that 'to know we know nothing', that heaven and hell resided in his own heart, that all monsters were within him and that only in his dreams lay real understanding, beyond reason. He dismissed book knowledge and didn't want us to 'waste our days in the blind pursuit of learning'. Hudson had turned from White to Browne, from facts to mystical debates.

Then, after his arrival in England, Hudson came up against ignorance of White. He recalled meeting a naturalist on the South Downs who admitted he had never heard of White of Selborne. Hudson's surprise that White was not universally known 'filled me with astonishment and even humiliation'. He had made an immigrant's mistake of assuming that everybody was familiar with the dated White. And that if they hadn't read him, they nevertheless would have absorbed his thinking. Thus reading on his own, out of context with his generation in England, began a life process of working out his 'own salvation with fear and trembling', while all the time thoughts of death pursued him.

Hudson was puzzled why his farm had the only library in the area. Over sixty years later, he recalled the order of the books he first read, starting with Charles Rollin's thirteen-volume *Ancient History* that opened a 'new wonderful world', as gripping as brother Edwin's bedtime yarns. He then read William Whiston's translation of *Josephus*, a *History of Christianity* in possibly eighteen volumes, and Augustine's *Confessions* and *City of God*. Leland and Carlyle followed and he was halfway through the six volumes of Gibbon's *The History of the Decline and Fall of the Roman Empire* when his father was ruined and they had to abandon Las Acacias. Elsewhere he remembered reading poetry by Eliza Cook, William Cullen Bryant, Bloomfield and Bunyan and Busching's six-volumed *A New System of Geography*. I defy anybody to imitate this anachronistic reading plan that excluded fiction as too frivolous. No wonder Hudson was so peculiar.

Hudson once told of a book-eating sheep, who would wait until left alone, walk into the shack, grab a book and rush out, pulling out the pages with its teeth. It tapped into his ambivalence about books. He both read and ripped up books.

But book learning was finally a delusion. He felt you had to observe and reflect for yourself. A deeper education for the growing Hudson was nature in the wild. Hudson was given the freedom to be and to observe. Once ill, and turning to books, he grasped what he had lost. No wonder *Far Away and Long Ago* ends on reaching eighteen years. Hudson contrasted his childhood experience of being immersed in nature with formal book learning: 'We are not *in* nature; we are out of her, having made our own conditions . . . what we are – artificial creatures.' He struggled to return to that unconscious bliss through roaming wild spots and through memory, but it wasn't the same. He felt stuck between the natural and the artificial worlds. He could not go back.

His oddity in England was furthered by his un-English attitudes to horses, dogs and snakes. First, some horse anecdotes. Dowager Ranee Margaret of Sarawak remembered when Hudson was in Wiltshire and suddenly leapt bareback onto a horse by pulling himself on by the mane. He galloped round and round a field on the bucking, kicking horse. He was seventy years old.

Once Hudson slipped into a daydream on top of a horse-drawn bus from Oxford Street going west. It was hardly moving, 'so elevating my umbrella,' Hudson recounted, 'I dealt the side of the omnibus a sounding blow, very much to the astonishment of my fellow-passengers', as if his umbrella was a horsewhip.

A last anecdote tells how, while out on a stroll, he bumped into Cunninghame Graham riding his horse Pampa on Hyde Park's Rotten Row. He went up to rider and horse, hugged the horse and burst into tears. Hudson didn't go mad, like Nietzsche did in Turin in 1889 on seeing a flogged horse, but he cried in grief for the horse and for himself. The fact is that Hudson, like any gaucho child on the pampas, was born on a horse and by six was wandering freely everywhere. There was no such entity as a gaucho without a horse. From early on, the horse meant freedom: no helmet, no fences, no rights of way, no bridal paths. Hudson reckoned that horse-riding was the 'nearest approach to bird-life' he knew, the closest to flying a man could achieve, a Pegasus reality.

In 'Horse and Man' in *The Naturalist in La Plata*, 1892, Hudson defended riding as 'exhilarating'. The horse gauges the ground so that the rider's mind is free. Riding was Hudson's clearest mode of

Florián Paueke, S. J., *Horsemen Fishing*, 1749.

thinking. From early infancy he rode so that 'we come' (meaning 'we gauchos') 'to look on man as a parasitical creature, fitted by nature to occupy the back of a horse'. Hudson said that the bow-legged gaucho, who waddles with toes inwards and is often accused of being lazy, when on horseback 'is of all men most active'. The gaucho without a horse has had his feet cut off, Hudson added, 'to use his own figurative language'. He belittled Darwin, again, for mocking a gaucho who claimed he was too poor to work but had failed to grasp that he meant that his horses had been stolen. In his day, even beggars begged from horseback, fishermen fished from horses and gauchos took mass on horseback.

Hudson also esteemed a horse for its emotional intelligence and how a horse 'keenly anticipates in our pleasures'. He tells a tale of when he fell in love with a frisky horse, aged thirteen (around 1854). His father gave him the money to buy it from a vagabond who gambled. Despite its seeming untameableness, this horse never threw Hudson. He thought it was because he didn't use a whip. Once, at a cattle branding, this horse watched gauchos charging down branded bulls for fun. Suddenly a bull that had escaped stirred Hudson's horse into action and he charged the bull and knocked it down. Hudson still recalled the gauchos applauding him, even though he had done nothing. It was his horse. He called this horse 'Picaso' (meaning a black one) and it was his until it died.

After a thistle year, when he could not ride, came the levelling wind. Old man Hudson in 1916 could still recall the 'ghost of a vanished thrill' of galloping off and hearing the horse's hard hooves 'crushing the hollow desiccated stalks'. Hudson remembered a horse left behind by a young officer who had fought guerrilla wars in Uruguay (and whose stories of the Banda Oriental were a source for *The Purple Land*). He asked Hudson's father to look after this favourite horse as he had to fight in the north. Nine years passed and he never returned to claim it. It was a dark brown horse with a long mane, but lean and old. The Hudson family all learnt to ride with this horse named Zango. It was killed in a hailstorm.

In general the gauchos and Indians owned many horses and changed to a fresh horse to continue covering long distances. Horses were cheap. Still, Hudson found horses tamed by Indians more docile. It understood its master with the slightest touch of the hand. If the gaucho sleeps on his horse, the Indian die on it. Hudson related a horse anecdote about a deserter from Rosas's Army named Santa Ana, who was a neighbour and who had escaped being caught and shot (as all deserters were). He slept rough, but when his horse heard horses in the distance it would shake the sleeping man by pulling at his poncho. The man then would hide in the dense reed beds. Hudson met him.

The gaucho horse, called a *criollo*, couldn't be more different from the English hunter. Like its gaucho rider, the horse showed endurance, was quick in action and stoic. The *criollo* was not a showy creature and never wasted its energy. Its sense of smell was astounding. It could smell rain fifty miles off and would rush off to drink. On the frontier, these same horses showed great terror. They knew when the raiding Indians, anointed in their rancid mare's fat, were coming and would scatter. Hudson himself was once on a horse that smelt Indians and bolted.

Hudson ends his chapter on pampas horses by evoking his favourite activity, riding at night, lying flat on his back, his feet pressed against the horse's neck and staring at the stars – a method of riding that was 'impracticable in England'. Throughout his work, Hudson differentiated the Arabian or *criollo* horse of his Argentine past from English horses.

Hudson told an anecdote about a childhood horse called Moro, due to its dark colour, that had been acquired from a neighbouring gaucho. During wet weather Moro would escape back to its home. It would be fetched and would remain until more wet weather. A horse's homing-instinct was never forgotten. During a cold rain spell at his second home, the horse was reminded of home, 'the wide green plain bathed in everlasting genial sunshine', and fled back, just as Hudson did through memory from his second home, England.

Pedro de Mendoza reached what would become Buenos Aires on 2 February 1535, with seventy horses on board. He built a primitive fort and settled. There were no horses in the New World (though there are fossil remains). After three years of famine and hostility, the Spaniards abandoned their mud fort and left around five mares and two stallions, though figures vary. Over the centuries what Luis Franco called Arab-Berbers multiplied on the grass lands, flat like the deserts they came from before conquering Spain with their Moorish masters in 711.

In 1853 William MacCann saw such an immense herd of wild horses that it took eight hours for them to trot past him. Before him, in 1744, Father Falkner, an English Jesuit who lived for over forty years on the plains, saw a herd of wild horses, going upwind, that took a fortnight to pass.

The Indians were the first to adopt the horse, riding on sheepskin saddles and developing incredible skills that were transferred to the gauchos, themselves mixed with Indian blood, so that horse riding became a necessity, a food, capital and a myth of identity.

What's now understandable is that Hudson *refused* to ride in England. He was no William Cobbett riding rurally, hadn't the means and didn't want to confuse what he meant by horse riding with fox hunting or parading on Rotten Row in Hyde Park with Cunninghame Graham. Walking and bike-riding were more attuned to freedom in England than riding. Hudson had arrived in England just as the horse was being superseded, after thousands of years as the best and fastest means of transport, by the motor car and steam train.

Biographer Luis Franco, in his *Hudson a caballo* (1956), touched on two further points. Round the kitchen fires in all the gaucho huts, horse-talk dominated. Whinnying and the sound of hooves on the earth define horse music. There's nothing more satisfying than this thud of hooves as a stallion and his mares gallop past you. Also, the horse is a strange beast. Hudson recalls the horse's twitching power to shake off dust, flies and even the rider. A man riding bareback exclaimed: 'It's like riding on an earth-quake'. Hudson had no experience of quakes and thought it more like an electric shock. After having been ridden, and had its saddle unclipped, the horse rolls on its back, free at last. When Hudson reached England, the horse was losing its everyday, working companionship, and was becoming a luxury he couldn't afford.

English dogs were also compared to the animals he knew in his early years. Hudson hated pets, or 'flattering parasites' as he called them. Yet, working as a gaucho, dogs were essential. A shepherd doesn't exist without his sheepdog. The pampas in his day also had lethal packs of dogs gone wild. Every time Hudson approached a ranch or farm, dogs would rush at him barking. One dog, nicknamed Pichincho (meaning doggie in Spanish), became a playmate. But it was never a pet or kept indoors or mollycoddled.

Still, the death of his old family dog Caesar in 1847 shook young Hudson. He did all he could on the dog's last day, warming him with rugs, and offering drink and food. After he died Mr Trigg the tutor suggested a burial service. A grave was dug among the peach trees. Then he gave a spontaneous speech, ending: 'We die like old Caesar, and are put into the ground and have the earth shovelled over us.' Hudson was pierced to the heart. He had awoken to the fact that death existed. Personal death shocked him, despite being in a country where 'battle, murder and sudden death' prevailed.

Another dog story emerged from *Idle Days in Patagonia*. A self-inflicted knee injury (of which more later) forced Hudson to pay attention to more domestic matters. At an English *estancia*, he befriended a retriever with an extremely curly coat called Major. He was old and blind, but implored Hudson to take him hunting. Hudson came across a flock of flamingos in a lagoon. Creeping towards them, Hudson, with increasing excitement, picked out a perfect specimen to shoot. He pressed the trigger and it fell, but he couldn't wade out over dangerous mud to fetch it. So he sent Major who managed to bring back the large cock flamingo. But when he went hunting for geese, Major acted differently. Cold boiled goose and coffee, without bread, was Hudson's daily Patagonian breakfast. Hudson shot several geese, but Major refused to bring them back; instead he deposited them all on an island and then pulled them to pieces. He then swam back to Hudson without one goose. Hudson called this a 'sudden reversion to the irresponsible wild dog', much as a human could suddenly regress back to a savage. Major had belonged to the Earl of Zetland but had taken to killing sheep and had been sent to Patagonia to save his life.

In 1919 Hudson finally collected 'The Great Dog-Superstition' in his *The Book of a Naturalist*; it had been first published in 1889 in *Macmillan's Magazine*. This elegant invective had irritated dog lovers or canophiles, as he called them, especially Frances Power Cobbe. She insulted Hudson as 'worse than a vivisectionist', as he recalled in 1919. Cobbe, the overweight author of an essay 'The Consciousness of Dogs', admired by Darwin and esteemed by Hudson, was a founder member of the anti-vivisection society, particularly aimed at stopping vivisection on living dogs at University College London.

Hudson published his piece anonymously. His editor Mowbray Morris received the essay with a 'painful shock', afraid Hudson would disgust many readers. The fact is Hudson despised lap dogs and the 'mawkish dog-sentiment' of their female owners. The 'Dog-Superstition' was that dogs were superior, more intelligent than other mammals. Hudson compiled a two-page list of equally intelligent mammals from squirrels to vizcachas, a rodent found in South America. For him, the lap-dog was still the jackal of disgusting habits, superficially domesticated. Hudson had learnt on the pampas that dogs, essential for sheep-farming, should be kept outdoors, like horses, pigs and cows: 'We may wash him daily with many waters, but the jackal taint remains.'

Hudson, shocked at his readers' response, didn't develop his invective. He wanted to belong on the same footing as agitators and anti-vivisectors like Cobbe. However, he had confronted Walter de la Mare's truism: the 'Englishman's idol is a dog'. In a letter to Henry Salt, just before he died in 1921, Hudson bemoaned dog shit on pavements and counted the flies, which later would invade dining rooms and infect humans. Someday, he warned, sanitary authorities will look to this, and they have.

From an early age he had observed snakes, but something about seeing a living snake always startled him (and all of us) – they are reptiles that inspire a biblical and evolutionary revulsion. Hudson's first sight of the deadly *víbora de la cruz*, a pit viper snake, caused him a thrill of horror. An elder brother had dug up an opossum, with this snake curled up by its side. It was immediately killed, as all snakes were. Later, Hudson learnt how to pick them up, as he did adders in England.

He identified at least eight kinds on the pampas. If it wasn't for birds, Hudson wanted to become an ophiologist and told A. R. Wallace in a letter of 24 October 1904 that he regretted not having written a book about serpents. He penned two memorable chapters on them in *Far Away and Long Ago*. In the rented house, Las Acacias, Hudson would listen to snakes talking under the floorboards:

> A hissing conversation it is true, but not unmodulated or without considerable variety in it; a long sibilation would be followed by a distinctly-heard ticking sounds, as of a husky-ticking clock, and after ten or twenty or thirty ticks another hiss, like a long expiring sigh, sometimes with a tremble in it as of a dry leaf swiftly vibrating in the wind. No sooner had one ceased than another would begin; and so it would go on, demand and response, strophe and antistrophe; and at intervals several voices would unite in a kind of low mysterious chorus, death-watch and flutter and hiss; while I, lying awake in my bed, listened and trembled.

Here, reverting to being a child in the dark, he revelled in his fascination for wildlife. Snakes wake up awe, that trembling of the sacred.

However, the best snake anecdote concerns what the boy thought was a new species. He first spotted this large black snake in a wasteland behind Las Acacias. He gazed, 'thrilled with terror', as the snake slithered past him like a 'coal-black current . . . of some element as quicksilver moving in a rope-like stream'. The boy stalked this snake, trembling with fear, until by chance the snake crossed his instep. He shook in 'spasms of terror'. Nobody had ever heard of a black snake. It must have seemed like a waking-dream, so vivid with children when struck dumb with fear. Later Hudson, by now grown into a naturalist, solved the mystery. Melanism had turned a greenish-grey snake into a black one, though it was still unusual, and Hudson took fourteen years to identify it as a *Philodryas scotti*.

Hudson remembered another incident before leaving Argentina 'as if it had happened yesterday'. To everyone's horror, a Mrs Blake had saved a snake from being killed. She had merely said: 'Why should you kill it?' Her example taught Hudson one of his primal insights about life: 'whether it might not be better to spare than to kill, better not only for the animal, but for the soul'. This Mrs Blake was a typical lower middle-class Englishwoman 'who read no books and conversed with considerable misuse of the aspirate'. The Blakes appeared to stand behind a high wall and refused to mix with their fellows. They had built up a considerable fortune through hard work, but isolation drove Mr Blake to the bottle. Mrs Blake would often drop round and sing 'Home Sweet Home'.

Hudson's proposed book on snakes would start with the Druidic adder and adder-stone, but he always kept in mind that an adder was not an adder if preserved in a jar in a museum. In 'Hints to adder-seekers', he knew that only living adders told their story. You must not kill to know. It was Linda Gardiner's first law of the naturalist – do not kill – which became his dogma later, in England.

In 1870, in a letter, Hudson said he was 'fond of gunning' ducks and snipe. He had no qualms in shooting birds, scraping away the flesh and preserving the skins. From a manual he learned how to dissect and preserve, and sent nearly 500 shot bird-skins to the Smithsonian and Zoological Society, and must have shot countless more. In fact, he never set off on horse to bird spot without his shotgun. The fact is that to see a bird close up he had to shoot it. His first gun was memorable – early life is the saga of 'first things' – a single-barrel fowling-piece. It was flintlocked, of an iron-black wood with silver mountings and taller than he was. It became part of him, with its own intelligence. He copied brother Edwin in shooting duck for the family table. He would crawl up as close as possible and then shoot. The pampas, with their lagoons, abounded in water birds and still do. As you approach any 'laguna', clouds of ducks and waders nosily take off and fly around in circles.

Around the time of the battle of Caseros, 1852, the Hudsons ran out of bullets and feared being ransacked by the defeated army drifting past, so Edwin and his brothers made their own bullets from melted lead, and cleaned their armoury of old guns. But a local army kept the defeated marauders away.

Then when Hudson arrived in England in 1874, something happened: he stopped shooting and railed against it. He came to loathe pheasant shooting as much as the class of men who shot. The binoculars Morley Roberts persuaded him to buy satisfied his need to get close to a bird. He wrote 'of all man's inventions, this is to me the most like a divine gift', and never was without a pair ever again. Armed with binoculars, he developed a resistance to killing. In 1908 he wrote to Wilfrid Blunt bemoaning that in England the fowling passion was strongest and everyone had a gun and dreamt of shooting his beloved wild geese. He thought that immigration in Argentina had introduced another kind of bird-killer, the men who randomly shot everything that flew overhead, mainly Italians, but also Spaniards and Maltese. He became pessimistic about the future of birds with such callous killing. Hudson's choice of not shooting went against the grain.

Hudson's world was demarcated by the pampas and wind. The flatness of the pampas, with few trees in sight and the odd ombú denoting some habitation, shocked all travellers to the River Plate. Such a bowl of blue sky, with sunsets stretching across the whole arc, and nights studded with glittering stars are still unique. It was called 'el desierto' because it was deserted. It suggested 'infinity' because there was nothing between you and the low horizon.

Over my lifetime, the horizon has changed. There are many clumps of trees, with eucalyptus and acacia predominant, but they are rich with bird activities and song. Hudson wouldn't have believed this change from solitary ombúes to woods. Cattle have become scarce as the price of beef is controlled politically, but this could revert. There are feedlots and agriculture where a generation before it was deemed impossible to grow wheat or oats. However, there is still the hypnosis of flatness and size. It's hard to imagine an immense field of flowering mint or where a 'loma' or hill is about a foot higher than the rest. Best of all for me are the approaching storms where you see and hear the complete build-up of black clouds and forked lightning in the distance. And when it pours with rain and countless flat lakes spring up and the roads become impassable, the land can still regress to its primeval state of grassy emptiness.

José Hernández did not describe this ocean-like flatness as he took it for granted in the first part of his protest poem *Martín Fierro*. It took a foreigner to notice the differences with Europe's rolling hills and mountains. Sir Francis Bond Head, first to bring to European readers like Darwin a description of the empty pampas in 1826, noted that 'our ignorance of the country' was immense. He evoked its extent, with the smell of clover and 'the sight of the wild cattle grazing in full liberty on such a pasture' as very beautiful. Then Head added that the state of the unploughed pampas is as it was since 'the first year of its creation'.

In the 1990s it was accepted that wheat and corn could be harvested at Las Tres Marías, my wife's family *estancia*. Maybe some of the land had been ploughed to reseed a more nutritious grass like *festuca*. But I remember the shock of thinking that the tractor cut into turf that had *never* known the plough. The best evocation of the pampas is that of Drieu La Rochelle. He was editor of the Parisian *Nouvelle Revue Française* and lover of Victoria Ocampo. He'd quipped that the pampas were a 'horizontal vertigo'. Hudson knew that his experiences in the wide-empty, acoustic pampas made him unique in England.

In the opening chapter to his *The Naturalist in La Plata* (1892), he described the pampas as his 'parish of Selborne'. He prowled its natural history like White did his parish. The Quechua word *'pampa'* meant open country or plain (but he doesn't say it was first applied to high Andean valleys) and was anglicized in the plural, pampas. When Hudson was there, before the Argentine government's ethnic cleansing in 1879, the actual Argentine republic ended in an arc of 200 miles from the capital.

Hudson found Darwin's explanation for the pampas's lack of trees (extreme winds) unsatisfactory, given that the ubiquitous euca-lyptus had taken root, this 'noble tree' gaining heights and luxuriance of foliage unknown in Australia, from where Sarmiento first imported its seeds in 1858. However, the grassy, humid 'desert' is home to the stately pampas grass, now common in suburban gardens. Hudson raved about this reedy grass in its proper site:

> I have ridden through many leagues of this grass with the feathery spikes high as my head, and often higher, It would be impossible for me to give anything like an adequate idea of the exquisite loveliness . . . of this queen of the grasses, the chief glory of the solitary pampa.

The garden variety is bedraggled and ugly but, as I have seen, in late summer and at sunset and in moonlight on the pampas, this grass takes on unbelievable hues in the plumes.

Hudson referred to March 1874, just before leaving Argentina, and the last time he caught sight of pampas grass. He had been travelling with a companion all day through this matchless grass when he turned round sharply and saw a party of mounted, possibly hostile Indians:

at the very moment we saw them their animals came to a dead halt, and at the same instant the five riders leaped up, and stood erect on their horses' backs. Satisfied that they had no intention of attacking us, and were only looking out for strayed horses, we continued watching them.

These silent, gazing bronze men with dark copper skin and long black hair amongst the plumes of the pampas grass echoed in his memory. They ignored Hudson and his companion, in their own world. It's like a painting.

Hudson 'loved the wind'. In an early letter of September 1866 to Spencer Fullerton Baird of the Smithsonian, Hudson evoked the 'cold South winds of winter' and the 'scorching North winds that blow incessantly in summer'. Most people hate wind. Here is an example of Hudson's quirkiness, until you grasp that wind is another way of saying pampas. The treeless plains, with solitary ombúes, ensure that nothing impedes the wind. The wind can blow for days on end and is so familiar it has names: the cold *pampero* from Patagonia and the warm *sudestada* off the River Plate. On an island like Britain, with hills, valleys and woods, wind is not so predominant, as it might be at sea.

Hudson's yearning for wind was another form of homesickness. He enjoyed roaming the Downs because there wind blew into his face, and reminded him of home where ostriches opened out their wings and sailed with the wind. The wind could herald incredible storms, visible in the wide-open sky as they approached. Wind brought Hudson 'face to face' with one of the expressions of wild nature. In England, Hudson missed this endlessly howling wind.

Here's Hudson's wind therapy: to lie or sit for an hour 'listening to the wind'. I took his advice. It's very restorative. You listen and think of nothing, 'simply living in the sound of the wind, that strange feeling which is unrelated to anything that concerns us'. Some inexplicable intelligence in nature then takes over the mind. But I couldn't keep it up. The fierce pampas wind began to scare me. It could rip off roofs. It could turn into a tornado or push you over. It's almost too elemental and that's the difference between Hudson and me: that he sought to lose himself in elemental forces.

Hudson recounted how he would lie awake at night at Land's End, fascinated by winter storms. The wind revived ancestral fears. He had located a *wind-sense* (his italics). Listening to wind restores the lost, natural man. In a nostalgic early poem, 'In the Wilderness', Hudson wrote: 'There winds would sing to me'. He recalled how in 1860, aged nineteen, he was asked to supervise the shearing of sheep at a dusty and hot *estancia*. In three weeks some 40,000 sheep were sheared (men could do up to 120 sheep a day; women up to fifty). Hudson was so bored at night that he read a hagiography of San Juan Gualberto, the only book in the hut, where a hostile wind lifted the gown over the head of a woman who had mocked the saint. It reminded me of Federico García Lorca's gypsy ballad 'Preciosa y el aire' where the wind is personified as a priapic San Cristobalón, who wants to lift up the gypsy girl Preciosa's skirt.

Wind meant most to Hudson when on horseback for weeks at a stretch from dawn to sunset, like any gaucho. The most exciting time to ride was during strong winds for that was when he could really think. Was it more oxygen in his brain? Was it the wind's vibration on his galloping body? Whatever, his mind was like birds in flight, as if the wind blew through him and left him with a 'perfect freedom of mind'. He cannot explain this feeling, except that 'vestiges of ancient outlived impulses, senses, instincts, faculties' are stirred up by wind. He even felt that two experiences he had of telepathy, seeing a clear face in front of him, were due to wind force.

Chapter 5

Growing Up on the Pampas

Turning back to Hudson's Argentinian years led me to try to resolve a puzzle in *Far Away and Long Ago*. One special tree on Hudson's small farm was the *paraíso*, a tree with poisonous leaves transplanted from Entre Ríos state in northeast Argentina. But surprisingly it originated in Tibet. It made me wonder about the immigration of plants. At the Hudson museum, the team are planning to remove all foreign species and promote native ones by giving them away. But you cannot stop plant immigrants. If that was a national policy, what would happen in England to the potato, the tomato and the corn on the cob? And these are just three New World species.

Are these immigrant trees called 'paradise' trees in Argentina because of their shade? Imported by Domingo Sarmiento as seeds, they were considered national enough to once line Buenos Aires's main Plaza de la Victoria (later, Plaza de Mayo), before being uprooted and replaced with palm trees, which were also ripped out.

At Las Tres Marías there's a complex of wooden fences for sepa-
rating cattle that's wonderfully aged with moss and colourful lichens.
Cattle are herded into a corral and then pushed into single file by
dogs and yelling *peones*. The cattle can be stopped individually by
shutting a sliding wooden contraption; they are then injected, have
their eyes treated for sun-cancer and swarms of biting flies flitted. It's
hard work, and it's sited in an isolated spot on the pampas. However,
there are four *paraíso* trees growing there, providing shade during the
blistering summers. They were planted from seedlings from a tree in
Entre Ríos and look as if they've always been there on the pampas.

Hudson recalled this *paraíso* tree's fragrance, stronger than
orange or lime, which reminded him of his exile:

> I often stand, in memory, in the shade of its light loose feath-
> ery foliage, drinking in the divine fragrance of its dim purple
> flowers, until I grow sick with longing, and being so far
> removed from it feel that I am indeed an exile and stranger in
> a strange land.

Behind the Bible quotation from Moses's exile in Exodus 2:22 ('stranger in a strange land'), reminding us of Hudson's deep familiarity with the Bible, the memory of this tree hurled him back to the *monte*, or wood, on the pampas. Smells, he speculated, do not register impressions in the brain like sights, so an odour is always a novelty. You cannot *think* of a fragrance, you have to let it hit your nose as the Madeleine hits Proust's palate. Hudson's Strangeland mutates easily into England.

But now to the puzzle. In Hudson's time, close to the main building at Los Veinte-cinco Ombúes stood a huge tree simply called 'the tree'. Neighbours said it was unique, the only one in the world. All Hudson could recall was that it had white bark, long white thorns and dark-green leaves. In November it was covered with tassels of minute, wax-like flowers that were strongly scented and straw-coloured. There was no tree book on native trees to consult while he was out there so when writing his memoirs, aged seventy-seven, he was obliged to leave it as a mythical tree.

The resident botanist at the Museo Hudson, taking me on a tour of trees and seeds, convinced me it was a tarumá (Citharexylum montevidense), a native tree. It has been suggested as one of three possible trees by a previous biographer, but as biographers are not usually good botanists, I needed to know. The tarumá, sounding so Guaraní, still stands. Was it the actual tree that Hudson knew? The Museo team think so. But how old is it? Was this the tree where, suspended from his wrists on a branch, a young black slave, who had fallen in love with his mistress, was scourged to death by fellow slaves? He was buried behind one of the ombú trees to become the house ghost to young Hudson's childhood. Not knowing the tree's popular name exemplifies Hudson's mytho-poetic, boy's view of the world, despite having written about it in his old age.

It's a limited boy's view we get throughout *Far Away and Long Ago*. Hudson's succinct portraits, through his old man/child's eyes, of neighbours at Los Veinte-cinco Ombúes and Las Acacias, are brilliant and function as crucial social history of Argentina during his years there. These neighbours form the human backdrop to that moment before immigration when rural Argentina still catered for the individuality and difference that also defined Hudson himself. All are eccentrics because civic society did not exist.

There was old cigar-smoking Doña Pascuala, with her authoritative voice and masterful manner. She interrogated Hudson about his family. Did they really drink coffee every day? There was the Barboza family, run by the father with eagle-like eyes under bushy eyebrows and immense crow-black beard that spread out over his chest. He always carried the gaucho knife and was a famous fighter, as already noted. There was the Casa Antigua run by educated Englishman George Royd and his fat Spanish wife, with very white skin and raven-black hair, always in her chair with a fan. He committed suicide.

There was the *estancia* La Tapera owned by Don Gregorio Gándara, with a fat indolent wife who kept parrots. He had staring, bulging eyes like a toad and a grotesque laugh, as if having a fit or screaming like a fox. He bred piebald horses and spent his time with the thousands he had. The Hudsons had rented Las Acacias from a Gándara. Today, there's a village called Gándara.

There was Don Anastasio Buenvida at the *estancia* Cañada Seca, a bachelor dandy who spent hours every morning preparing his ringlets of long hair. He was devoted to pigs. When Hudson was about eight, one of his red sows charged him.

And there was the patriarch Don Evaristo Peñalva, with his six wives. He was a kind of doctor to the gauchos. He cured shingles by writing 'In the Name of the Lord' near the wound, and then rubbing a live toad that exuded a poisonous milky substance over it. In 1864, on his way home from Azul, Hudson found this patriarch in his second *estancia*, La Paja Brava, reduced to one wife and a disgraced daughter who had previously run off with a lover.

No one has traced these families to see if they really existed, but this extraordinary gallery, vividly described as if seen for the first time by Hudson-the-boy, defies oblivion. Were they so engrained in his memory that they leapt to the page in words or did he reinvent them as counters to boring English neighbours, as if the classless pampas bred only eccentrics, including Hudson?

Far Away and Long Ago, in which Hudson recreated those neighbours, emerged during illness in England. In 1915 he was ill in Cornwall and his wife Emily was even more ill in Worthing. As he vividly recounted, he was visited by emotional images from growing up wild in Quilmes and near Chascomús, and started to write his autobiography.

When he was at St Michael's Hospital in Hayle, Cornwall, he was seventy-four years old, being tended by nuns, and feeding birds from his windowsill. Through his correspondence with his publisher Alfred Knopf (Hudson never employed a literary agent and dealt directly with Knopf) we see the autobiography changing titles, from 'Long Ago and Far Away' in 1916 to 'The Book of Early Memories' in 1918.

He told Galsworthy that this book was more or less finished in January 1916, but he told Knopf that he often had to stop writing due to bouts of illness. In 1917, according to a letter to Knopf, it remained unfinished.

He defined the book's scope to Galsworthy: 'All the interesting part of my life ended when I ceased to be a boy and my autobiography end [*sic*] at fifteen', which, as already noted, is untrue, as it ended with the death of his mother in 1859 when he was eighteen years old. Parts of the manuscript have survived and are held at the RSPB library. I enjoyed reading them again, but learned nothing new. I did find it moving that Linda Gardiner typed at least two of the chapters. However, a myth has grown about this visionary basis

W. H. Hudson c1915.

that is not exact and it was promoted by Hudson himself. He wanted to be a writer of visionary import, beyond the usual literary fiddle. His book was to be a 'clear and continuous vision of the past'. And he suppressed or forgot that he had attempted it several times and that writing is about words, not only vision. He felt like William Blake being dictated to by some force. He was a visionary. But he had tried his memoirs many times before.

In fact, the first attempt at an autobiography appeared in 1886 as 'Far' in the *Gentleman's Magazine*. It would be the basis of the opening chapter of *Far Away and Long Ago*. The style is different, more speculative. He tells us he'll write about a region '*so far* away from England', his italics, and so unfamiliar. So far back that he saw his childhood as another's, but he was 'exceedingly happy' then. This state of childhood bliss is an essential clue to the later man. He never boasted of his bliss, hardly alluded to it, and was not nostalgic.

The rest of this early story is familiar, the giant ombú in which the children built a treehouse, the flat plain on one side and

Ms page from *Far Away and Long Ago*,
held at the RSPB library.

undulating down to a stream on the other, the story of the young black strung up on the tree, visits from the Hermit and his mother, sitting out of doors at sunset, with an enigmatic smile on her face.

In June 1912, twenty-six years later – a long incubation – he published an essay, 'A Memory of the Ancient Time', in the *English Review*. It's wedged between poems by Ezra Pound and D. H. Lawrence and a story by E. M. Forster. By now, thanks to the efforts of Ford Madox Hueffer (who became Ford Madox Ford), he's with the modernists. This version doesn't differ much from the opening chapter of *Far Away and Long Ago*, except we get more details – he was five when he left his birthplace, the month was a cold morning in June, it was a longer journey, from sunrise to sunset, his youngest sister was born in the new home and the ditch or foss round the house (guarding from marauding Indians) was, exact measurements, forty foot wide, eighteen deep, with water and 'a paradise of rats'. In fact, rats terrified the boys and 'ran over beds' at night while they yelled in alarm.

The roving tutor, hired by the month, was named as Mr Triggle in 1912 rather than Mr Trigg. He was a man of 'terrible sternness', a nomad with a saddle bag and horse. He pinched the boys, until he lost his temper and beat them with a rawhide horsewhip. They listened to two hours of Dickens read aloud each evening.

Hudson characterized Argentina as a 'land of battle, murder and sudden death' and there was 'seldom a day on which I did not see something killed'. He revisited the scene where he found, tethered to a pole in the barn, a murderer, adding 'murder was a common word in those days', to shock his English reader. A Rothenstein portrait of Hudson accompanied the piece.

Hudson reworked it five years later in a final version, dealing with how memory works. But I'm sure he didn't have it to hand as he wrote, as the details are different and the tutor's name changes from Triggle to Trigg.

In 1916, in a letter to the Ranee of Sarawak, looking back on his life he found it strange, almost incredible, 'from the childish days under the giant ombu trees watching the men driving the herd of bellowing cattle'. He joked that well and truly did Cunninghame Graham say that he was no use, that 'he thinks more of a willow wren' than any of this friends. But the immense distance of his own past when he was a happy, free child had only increased his loneliness. He had to write it down as there was nobody with whom to share it.

Chronologically, Hudson's next description of his life in Argentina leaps from his pampas childhood to his aptly titled *Idle Days in Patagonia*, 1893. The exact dates for his Patagonian trip are not known, but they follow his father's death in 1868. In this book, he mentions arriving in summer and the shock of seeing snow for the first time. He took a 'tumbling, storm-vexed old steamer', shaped like a Viking longship, from the port of San Borombón down the long coast to Carmen de Patagonés. This outpost is not technically in Patagonia, but at the edge of the province of Buenos Aires. Beyond this town lay Indian territory.

I passed through Carmen when President Alfonsín was arguing to move the capital there. It would promote regional development and become Argentina's Brasilia. But his absurd desire remained stuck as an idea and nothing happened in a place where already nothing happened. Even the train I took no longer runs.

Hudson's old boat founded on rocks and was beached so that he and the other passengers had to wade ashore and walk some thirty miles to the Argentine outpost as most of Patagonia was still in Tehuelche hands. Carmen de Patagonés (and Viedma on the opposite bank), thirty miles up the Río Negro, was founded in colonial times in 1779 as Spain's most southern trading post, but it could only be reached by sea. It lies 937 kilometres from Buenos Aires and in 1870 held some 2,000 inhabitants. Hudson wrote about this shipwreck twenty-three years later in Paddington. Was he vying in prose with his friend Joseph Conrad? Was there a wreck at all? There's something in the intensity of the opening that makes me suspicious.

I was just dropping into a doze when a succession of bumps, accompanied by strange grating and grinding noises, and shuddering motions of the ship caused me to start again and rush to the cabin door. The night was still black and starless, with wind and rain, but for acres round us the sea was whiter than milk. I did not step out; close to me, half-way between my cabin door and the bulwarks, where our only boat was fastened, three of the sailors were standing together talking in low tones. 'We are lost,' I heard one say . . .

Then an English-born engineer appears, loaded pistol in hand, to prevent the men abandoning ship in 'that awful white surf'. By the publication of *Idle Days in Patagonia*, Hudson was already a published novelist, with plenty of manuscripts piling up at his side. No report of a ship foundering had been written about in the local press, so maybe his fiction bled into his memoir.

All the information we have is from Hudson, except that we now know the name of a young Englishman he met at Carmen, who had remained anonymous in *Idle Days in Patagonia*. Zoologist Emiliano MacDonagh, on a field trip to Carmen in 1932, named him as Ernesto Buckland, cousin to well-known ornithologist Frank Buckland in London. Ernesto died soon after this meeting.

What happened next to Hudson is the crux of the travel book. Hudson and Buckland rode up the Río Negro and stayed the night in a hut Buckland kept. While handling a loaded revolver, Hudson pressed the delicate trigger and shot himself in the left knee. Blood poured out. Buckland bound him up, rode back to Carmen to fetch a cart and brought Hudson to the local surgeon, who struggled to find the bullet. That was the end of Hudson's travels for a while. Worse than the wound, however, was the realization that he had passed the night alone in the hut with a *víbora de la Cruz* or pit viper. Hudson recuperated in the English Mission House all February (summer) as it took the local doctor a month of daily, painful attempts to find and extract the bullet.

George Musters named this 'doctor' as the Revd Dr Humble. Musters had arrived in Carmen in the same 1870, after a year-long, 1,700-mile trip across Patagonia, but never met Hudson. He was commander of a sloop, given a year's leave, and published his *At Home with the Patagonians* in 1871. He died in 1879, aged thirty-nine, and left his Bolivian wife a widow.

While recovering from his wound, Hudson stayed with a Ventura Crespo near Barranca de los Lobos. There was no reading matter except a prayer book in Spanish, which he read aloud to his host, who was illiterate. Lack of books forced Hudson to mingle with the locals, and listen. In early 1872 he sent all that he collected, from bird skins to skulls and arrowheads, by boat. It sank and he lost everything. It was far more painful than his knee wound, he wrote.

Biographer Luis Horacio Velázquez worked out that Hudson stayed in Patagonia from October 1870 to March 1872, though Hudson himself told Col. Lane Fox at the Pitt Rivers Museum in Oxford that he had spent a year in Patagonia from 1870–71.

Hudson's time, after his self-inflicted wound, riding alone up the Río Negro, was his happiest, cut off from city news and exploring an unknown territory with Félix de Azara's book on Paraguayan birds as sole guidebook. Azara (1746–1821) felt stuck and forgotten in the New World, delayed by futile bureaucracy, and became a naturalist to pass the time. He found himself in a vast, unknown continent, ignored by science and without books and conversation with equals, except for a parish priest in San Ignacio. He aimed to correct the French encyclopaedist Buffon, as much as the false myths invented by Jesuits. He wrote in a style where things matter more than words, against the 'novelists'. He was one of the first anthropologists, studying the indigenous ways. His book was an apt companion to Hudson, who felt the same. But his published notes were purely verbal, lacked drawings and offered confused taxonomy. In three volumes he noted down 448 bird species. I can see Hudson struggling to match what he shot with what Azara had annotated.

But as a late nineteenth-century explorer, Hudson's discoveries were meagre. He did find two new bird species, now named after him, but his ambitious dreams of finding Trapalanda, the gaucho mythic island of the dead, were curtailed, not only by shooting himself in the knee, but by lack of funds, and not being able to penetrate the desert for more than some 180 miles from Carmen de Patagonés. Unlike the naturalist

Alexander von Humboldt, he was not a millionaire, and unlike Darwin, his was not an officially funded trip. His stay was also brief when compared to earlier explorers like Darwin (five years), Humboldt (five years) or Henry Bates (eleven years). The fragmentary form of the book mirrors the lack of purpose. Even Hudson acknowledges that the chronological thread vanishes and the final chapters float away, isolated. But some incidents convey much about Hudson the man.

Idle Days in Patagonia appeared in 1893, partially illustrated by his Shoreham artist friend Alfred Hartley. Hudson had posed for these drawings. Hartley later painted a portrait of Hudson in 1889. It was exhibited in 1891 and then it was 'lost'. According to a press cutting I found in the RSPB archives, this 'Lost portrait of Hudson' was found on 19 December 1926. It had been hanging on the wall of Hudson's London home. Hartley, frail in health, a St Ives artist and member of the Royal Society of Etchers, travelled up to London to confirm it as his. It hung for a while in a literary agency off Trafalgar Square, and then became 'lost' again. I'd love to know who has it hanging on a wall.

Four moments in this Patagonian trip have particular appeal to me. The first concerns the large black leaf-cutting ant (*Atta sexdens*), an exclusively agricultural ant. The workers travel along paths carrying back a chunk of leaf that they let rot inside their underground nests. They feed off a fungus that grows from it. Hudson knew this species from childhood as these ants are ubiquitous on the pampas. Like any child, he had tried defying these ants by digging a hole and watching them fall in.

I've been fascinated by their tracks wherever I see them, even in Buenos Aires. My mother-in-law, a keen gardener, had seen these ants strip her rose bushes overnight. To stave off these marauders, she'd fit empty plastic water bottles around the stems (the ants would slip off), or would dig a small moat of engine oil, or place tufts of sheep fleece to get them tangled, or just pour boiling water

Alfred Hartley's portrait of
W. H. Hudson, 1889.

down into their huge underground nests, located miles away. But the relentless ants always won.

Hudson admired these leaf-cutting ants and was sure from observation that their stable civilization would continue to flourish on earth 'when our feverish dreams of progress has ceased to vex it'. It was another aspect of the 'timeless' side to nature that excited him. As a field naturalist, not even an ant escaped his eye, and an ant was antidote to our chaotic society and its myths of progress.

The second moment happened while writing his book from notes when a memory from childhood about gaucho eyesight and spectacles jumped back to him. Hudson found Humboldt wrong in assuming that Indians and gauchos had better eyesight than sedentary city-dwellers. He tells of an Englishman wearing glasses who was mocked by a gaucho. Was it to make him better looking? Or wiser? How can looking through a pane of glass make things clearer? Hudson was aware that sight got worse with age and simultaneously that sight for a gaucho on the pampas was his life. The Englishman replied that he'd been wearing glasses for twenty years and challenged the gaucho to try them out. 'What, see better with this thing?' Hudson translates. But the gaucho put them on gingerly and stared around him shouting in astonishment. The trees looked greener and a distant cart was red. Hudson later saw him galloping around with the new glasses on.

A third incident involves shooting and wounding a Magellanic eagle-owl. By the Río Negro, in a clump of trees, Hudson spied this large bird, hardened his heart and shot. It fell into a tangle of dark-green grass and stared at Hudson like a monster, with each feather standing on end and the tawny, barred tail spread out like a fan, its beak snapping incessantly like the clicking of a sewing machine. Hudson gazed with fascination at its furious eyes, quivering with yellow flames. He wrote: 'The dragon eyes of that Magellanic owl haunt me still, and when I remember them, the bird's death still weighs on my conscience.' Hudson managed to stuff this bird and it gathers dust in the La Plata museum.

But the core experience of Patagonia, and the fourth incident that particularly appealed to me, was to do with silence. Darwin once again became Hudson's antagonist. In a passage from the *Beagle* voyage Darwin confessed he couldn't guess why the deserts of Patagonia loomed before his eyes and possessed his soul. Could it be that these open spaces released his imagination? Darwin admitted that he'd failed to analyse this strange pleasure. Hudson had a try. Perhaps, in 1870, he travelled around Patagonia with Darwin's *Journal* or maybe he had Darwin's travel book open besides him on his desk while writing in 1892. But the chapter 'The Plains of Patagonia' is a sermon based on a Darwin text. Hudson, from Paddington, still saw Patagonia as if 'actually gazing at it', where a memory is restored to a direct experience.

One day during winter, seventy or eighty miles up the Río Negro, he was roaming on horseback, with gun and dog. He would ride aimlessly for hours at a stretch. Usually, riding was his best mode of thinking, but in Patagonia he was able to empty his mind of thought. Resting in a copse of trees, the silence was profound and perfect. No sounds, not even the muffled hoof strokes of his horse. Not even the sound of his horse munching grass. His mind was no longer a noisy, thinking machine, but had tapped into a state of *suspense* and *watchfulness* – his italics.

What was this state? He interpreted it as a sudden 'reversion to the primitive and wholly savage mental conditions'. He linked it to fear, quoting Thoreau's sensation on feeling a thrill of savage delight when he wanted to eat a squirrel raw or to children finding wild fruit and glutting on them. Patagonian solitude had dragged him *back*, again his italics, to an intense alertness, the mental state of the savage at the same level as the wild animals they kill or that kill them. This mystical at-oneness with wild nature prompted Hudson to view us humans as 'living sepulchres of a dead past' that slumbers on until the roar of a waterfall or the sound of rain or wind in leaves restores 'a memory of the ancient time and the bones rejoice and dance in the sepulchre'.

I love that image of your body as outer casing, with a spirit inside only rarely released. That is, Hudson had momentarily reverted to the savage in wild Patagonia nature, or, maybe more realistically, located it while writing it twenty-three years later in London. That was Darwin's pleasure. At the Royal Geographic Society in November 1978 Paul Theroux, sitting with Bruce Chatwin, called Hudson the 'serenest' man imaginable who never forgot Patagonia. His message in capital letters was TRY PATAGONIA. Unfortunately I wasn't there. But listening to its silence was Hudson's experience of a higher plane of existence.

I'm aware that Hudson never talked about himself directly, but always through others, through anecdotes. His inner life was reflected in what he saw and how he saw it. He was ever the curious man. So the 'idle' of *Idle Days in Patagonia* has nothing to do with the pleasures of being idle. For a good son of American puritans, idleness was a curse. His mind was never idle. Even when forced to be 'idle' by shooting himself, he still 'worked'. The book is about the irony of being made idle. So 'Try Patagonia' is linked to being made idle. His Patagonia was a short trip up the Río Negro, at the edge of unmapped lands. The vast deserts of Patagonia remained beyond his range. The content of his book contradicts its title.

Dr Burmeister introduced Hudson to the eleven-years-younger Francisco Pascasio Moreno. There weren't many scientists around at the time. Carefully excavating a ditch by the river Conchitas, Hudson had dug up indigenous remains, which he had donated in 1873 to Moreno's home museum. Soon after, in 1884, Moreno converted his findings into a properly housed collection when appointed first director of the palatial La Plata museum, which opened in 1888. He remained as director there until 1906. It has been cleaned of the political slogans and graffiti that desecrated the building when I was last there. It looks grand. In the high, domed hall, there's a statue of Moreno in white marble. I persuaded a guard to let us look into his closed office, left as it was in his time. I noticed in his library that he had Darwin in French, and on a large desk in front of an ornate fireplace were piled his honorary titles, including an 1895 diploma from the Royal Geographic Society, naming him a foreign member. A glass case had his field camera and traveller's knick-knacks. There was a lawn-mower in a corner. In the museum itself, amongst the stuffed birds, I found a Hudson corner with an oil portrait based on a New Forest photo of him, and a glass case with his favourite Argentine birds.

Moreno recorded Hudson's donation in a note published in Córdoba, 1874; he also mentions a human bone (metatarsal) belonging to the historic Querandí Indians that his friend V. H. Hudson (*sic*)

had found underwater in the Conchitas stream. Hudson, just back from Patagonia, told Moreno to explore the Río Negro from El Carmen. Moreno, by now one of Argentina's earliest anthropologists, followed his advice in 1873 and sailed to Carmen on the first of five trips to Patagonia. He was the first Argentinian to sail on some of the great Patagonian lakes, patriotically baptizing one Lago Argentino.

He wrote up his trip as *Viaje a la Patagonia Austral*, 1879. The adventure of his life was being captured by Mapuche Indian leader Sayhueque and then escaping on a raft down the Limay river.

In charge of the difficult border demarcation with Chile along the Andes volcanic mountain chain, Moreno came to London to settle the dispute, gave a lecture at the Royal Geographic Society in 1899 as an honorary member, was awarded the George IV Medal and earned his nickname 'Perito', meaning skilled in negotiating (having earned Argentina 42,000 square kilometres, held a plebiscite with Welsh farmers and diverted a river so that it ran to the Atlantic). I found a Chilean website accusing him today of land treachery. He lost his young wife

Portrait of W. H. Hudson in the La Plata Museum.

in 1897, dressed in black and never remarried, and wrote little else. He was given huge chunks of land around Bariloche and gave back 7,500 hectares to the state as the first national park Nahuel Huapi.

When I crossed the great, cold Nahuel Huapi lake from Bariloche, still a national park, the ferry passed Centinela Island. The captain, on hearing that he had aboard Moreno's great niece (my mother-in-law), announced it over the tannoy. Everybody clapped and he gave the three traditional hoots. We could not stop to visit the monument with its white cross where Moreno's remains, with wife and son, were reburied, wrapped in Indian ponchos, in 1934. A plaque states he was a 'scientist and pioneer'. In 1899 in London there's no record of Hudson attending Moreno's lecture or meeting him, but he could have.

What separated Hudson from Moreno was the latter's patriotic view of science. He was a derivative thinker, worshipping but not criticizing Darwin. He fully expressed his time's zeitgeist. Even his prose is conventional. But, above everything, Moreno proudly

El 'Perito' Francisco Pascasio Moreno (1852–1919).

contributed to his *patria* Argentina. There's a photo of young Moreno standing outside Hudson's Las Acacias on the road to Chascomús. It's hard to see if it really is Las Acacias, and there's no biography of this conservative, bitter man.

Despite friendship with Dr Burmeister and 'Perito' Moreno, two of Argentina's incipient scientists, Hudson yearned to belong to a greater community. He would have understood later accusations made that Moreno had abused indigenous burial practices by stealing urns and skulls and exhibiting them, or that he had exhibited captured wild Indians from the 1879 'desert' war of extermination, that ended indigenous hostility. In fact, his collection of mummified corpses has been returned to the indigenous communities. Moreno is best known today for the Perito Moreno glacier near Santa Cruz, not for his philanthropic or scientific deeds.

I was offered an oil portrait of this man by Andrea's aunt who had it from her aunt, Moreno's childhood friend, but I refused it. It should hang in the La Plata Museo, I said. I did accept a daguerreotype, which hangs in my study, and a teapot painted in black and white Mapuche stripes.

A great catastrophe hit Buenos Aires the year Hudson was in Patagonia and Perito Moreno's mother died of this yellow-fever plague of 1871. Hudson never wrote about it directly because he wasn't there, but it provided fuel for an early novel.

Most of Hudson's fiction is rooted in South America. Dennis Shrubsall, a meticulous biographer, calculated that a third of all his writing deals with the New World, and with fiction it's six out of eight books. In 1888 in London, Hudson serialized his novel *Ralph Herne* in eleven consecutive numbers of a magazine titled *Youth*. He had written it before *The Purple Land*, soon after arriving in England. Just before he died, he consented to it being published as a book by Knopf, selling it for $1,000 to raise money for the RSPB, though he never got round to writing a prologue. It appeared posthumously in 1923. Hudson himself found this early novel 'rather tedious and even twaddly'. He's right.

As Ruth Tomalin notes, this novel reverses Hudson's experience in that the young English doctor Ralph Herne arrived in Buenos Aires in 1871 to face a dreadful plague. Hudson told Alfred Knopf in a letter in 1921 that *Ralph Herne* was a 'true history', but that he did not witness the plague as he was still in Patagonia and only heard about it from newspapers and then on his return from his brothers and sisters (none of his siblings died).

In 1871, San Telmo, in the south of Buenos Aires, was hit by yellow fever induced by the female mosquito *Aedes Aegypti*, brought over with slaves from Africa. There was no cure and it was more lethal than malaria, primarily affecting urban conglomerations. In 1867–8 there had been cholera outbreaks, but the 1871 yellow-fever plague changed the city. A special train called the *'Tren Fúnebre'* took corpses to the newly inaugurated Chacarita cemetery, as the Recoleta cemetery overflowed with so many dead. Hudson's invented English doctor, exhausted from treating so many dying citizens, watches a sudden storm flood drag hundreds of coffins down the street.

The story goes that the densely peopled 'south' was abandoned by those who could move out to Flores, on higher, healthier ground or to what became the '*barrio norte*'. Hudson's doctor stayed in the crowded immigrant south where the disease raged. In fact, death was out of control: up to 583 people died on the worst day in April. Hudson called this the Black Day. Following the plague's devastation, the once-fashionable Sur was abandoned to immigrants. The plague, according to historian Luqui Lagleyze, killed between 13,614 and 14,467 *porteños* out of a population of 190,000.

Hudson's novel is a naïve love story but his was a real, sudden plague where hundreds of coffins were piled up above the ground, for even gravediggers died. Hudson's sister was asked if she would like to buy a coffin in preparation. People would drop dead in the street in 'unspeakable agonies' or uttering 'delirious shrieks', the rumbling carts carrying loads of corpses, with the desolate cry to bring out the dead, that was, he added, 'so long unheard in Europe'.

His secret feelings emerged in that comment. He was ignored in London as nobody knew or cared about his Argentina. Nobody in London would ever guess that Buenos Aires's streets were filled with hurrying, busy people, 'dressed very much as Londoners', as rapt in their business and with a constant, urban 'deafening noise'. So he emphasized what would surprise an English reader, just as the Spanish-American mind (his phrase) was puzzled by the game of cricket. When Herne was down in his love-luck, he felt a 'stranger in a strange land, crushed to the earth', exactly as Hudson had felt in the 1870s and 1880s in London. Herne could have predicted the plague, as 'over-crowded' Buenos Aires hadn't even 'the faintest apology for a drainage system'. He felt that this 1871 plague was as terrible as the Plague of London. E. R. (whoever he was) told

Hudson an anecdote about a rat in his garden, expecting an animal lover's answer. Instead, Hudson was angry, and told him to poison it. Hudson was still in touch with fears of the plague.

Hudson picked on the yellow-fever plague, even though he hadn't directly experienced it, because it deeply scarred everybody in his homeland. Hudson didn't care about art, though recalled seeing a painting while a boy in Buenos Aires, by an artist recently returned from Europe, of a lagoon with horses. It haunted him with such a 'fierce' pain that he yearned to become an artist. He also disliked his bird illustrators (preferring the living birds), but when art lets you see reality, it has a function, and Hudson's novel was prompted by an oil painting, *An Episode of the Yellow Fever Epidemic in Buenos Aires*, by the Uruguayan Juan Manuel Blanes (1830–1901) in the year of the plague.

The opening chapter of *Ralph Herne* has the narrator queuing in a gallery in Calle Diamante to study the painting of the plague. He doesn't give the painter's name, but recognizes the artist's Flemish fidelity to the facts. Two or three men in suits and top hats, maybe doctors, stand at an open door, staring at the ghastly scene inside. A dead man lies in a cot, 'his plague-blackened face distorted with its last look of agony'. His stiffened fingers still grip the coverlid. On the floor is his dead wife, black hair spread over the dusty tiles, her skin grey and her lips burnt black with the fire of pestilence. Next to her, her babe, unconscious of death, startled at the voices of the doctors and the sunlight. Though a fiction, this scene has since been corroborated by historian Zorraquín Becú, who said 10 pesos were charged and Blanes's picture hangs in the now-demolished first Teatro Colón on the Plaza de Mayo. The two men staring at the death scene were in fact well-known doctors who died of the plague, and there were long queues to view the painting.

Hudson's first published book in England was the double-decker *The Purple Land that England Lost* in 1885, but written over the decade

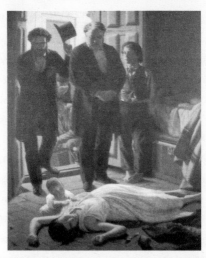

Juan Manuel Blanes, *La Fiebre Amarilla*, 1871.

after his arrival. With the second edition, shrunk to one volume in 1904, he excised half the title and it became *The Purple Land*. Hudson told Robert Cunninghame Graham that the character of Lamb wasn't autobiographical and that he had hardly been in Uruguay. He told Eliza Phillips, one of the RSPB founders, it was 'full of faults', and it does read like a boy's own adventure story. And yet, Borges found Lamb was like Hudson himself, a naturalist who included fellow humans in his gaze, without ever judging. Lamb's 'hospitality to receive all the vicissitudes of being, hostile or friendly' was Hudson himself.

After a decade in London on the breadline, but with an English wife, Hudson invented a fictive Englishman who did the opposite of what he'd done. It's the same inversion as in *Ralph Herne*. Richard Lamb stayed in South America – in this case, Uruguay – and slowly became South American. But in Hudson's struggle to become English, Lamb was an inverted alter ego. The years between the two editions are exactly the years of Hudson's assimilation (in 1900 he became a naturalized Englishman).

At the end of the romance – a picaresque litany of idiosyncratic country men and especially women with poetic names – Lamb reconsidered his English heritage. He began by pointing to a quality that Hudson suffered daily, the English contempt for foreigners: 'It is part of our unreasonable nature to distrust and dislike the things that are far removed and unfamiliar.' That he was an Argentine alien in London is the source of that perception. He hardly ever alluded to his national origins. Lamb then rid himself 'of these old English spectacles, framed in oak' to reveal what Uruguay gave him: 'the most perfect sympathy between me and the natives I mixed with' – such a different experience to the English caste system. He missed that 'wild, delightful flavour' of freedom. In his words: 'the sense of emancipation experienced in it by the wanderer from the Old World' where 'all men are absolutely free and equal'. No social classes and direct friction with nature became Hudson's topic. In Uruguay a republican freedom was 'hard to match anywhere else on the globe'. Even the Bedouin was not so free.

Lamb homed in on the absence of the class system:

> Here [in Uruguay] the lord of many leagues of land and herds unnumbered sits down to talk with the hired shepherd, a poor, bare-footed fellow in his smoky rancho and no class or caste difference divides them, no consciousness of their widely different positions chills the warm current of sympathy between two human hearts.

Lamb then reflected on England as a land 'with higher and lower classes, each with its innumerable hateful subdivisions – to one who aspires not to mingle with the class above him, yet who shudders at the slouching carriage and abject demeanour of the class beneath him!' To confirm that Lamb was Hudson, they fused for one moment: '"Do you know, Demetria . . . I intend writing a history of my wanderings in the Banda Oriental, and I will call my book *The Purple Land* . . .'"

Was it true that there were no social barriers in Hudson's Argentina of the 1840s to 1870s? In the long colonial period from the 1520s to the 1820s Spanish aristocrats chose Lima or Mexico City or Bogotá to return home rich with gold and silver bars from Potosí or Taxco or emeralds from countless mines. But in Argentina the lack of booty attracted a different colonial. For example, Juan Manuel de Rosas owned leagues of land, but mixed with his workers, rode his horses, played their games. At work, he was one of them. The same with Ricardo Güiraldes, millionaire, French-speaker and author of *Don Segundo Sombra*, 1927. On his *estancia* La Porteña, Güiraldes practised meditation but also dressed like his gauchos or *peones*, ate grilled meat by the fire, sang with his guitar and listened to their ghost stories, one of the men in a hard life of cattle herding.

Then, as Leopoldo Lugones pointed out, there was the abundance of meat. So much meat that most of a heifer, for example, was left to carrion birds and dogs. That everyone could eat as much as they wanted was a decisive factor in the making of an almost classless society. Even today this myth of abundance prevails. You can still order food in a restaurant where one dish serves for two.

Hudson was, for roughly twenty-four years, a sheep herder, out of doors, in wind and rain and always on horseback, sleeping rough, on equal footing with his bosses. He knew what he was writing about. He never sought his identity by dressing up as a Sunday gaucho. Ezequiel Martínez Estrada, essayist and poet, coined the right term for him 'gaucho from the inside'.

By 1874, this chumminess between owner and worker altered and gauchos became marginalized and persecuted. From when José Hernández published the first part of his outlaw's song *Martín Fierro* in 1872, land changed status with the ethnic cleansing of the Indians,

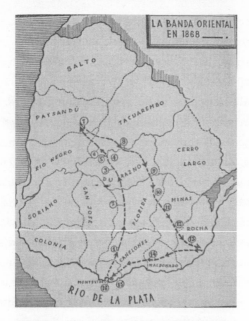

Luis Costa Herrera's Map of Hudson's Uruguay.

massive immigration, meat-freezing and the barbed-wiring of the great *estancias*. The freedom to roam on horseback ended and gauchos became *peones*, but Hudson had experienced the freedom his character Lamb lamented. Today there are only dressed-up Sunday gauchos, for the real, tough knife-fighter has vanished. Hudson had recorded a dying species.

The first edition of *The Purple Land that England Lost* hardly sold. Alicia Jurado claimed he began writing this novel in 1874, the year of his arrival in England, as part of a series titled *The History of the House of Lamb* that, according to Morley Roberts, he later destroyed. Roberts recalled that the manuscript was written on tiny notebook paper and stood about 2½ feet high. The second edition of 1904, cutting out *that England Lost*, did sell well. A letter from Edward Garnett guided Hudson: 'There are only two things to be done'; the first was to cancel the opening twenty pages and rewrite, for they were 'clumsy, gauche and jerky'. Start with Lamb on the mountain. Hudson carried out this change. Garnett also told Hudson to rewrite the last six pages of Chapter XXIX. The letter ended: 'I shouldn't touch another line in the book'. Hudson complied. T. E. Lawrence read it eleven times. Ernest Hemingway mocked his character Robert Cohn in *The Sun Also Rises,* who'd read it as 'splendid imaginary amorous adventures of a perfect English gentleman in an intensely romantic land, the scenery of which is very well described'.

There's been conjecture over whether Hudson covered the same Uruguayan ground as his protagonist. Uruguayan critic Luis Costa Herrera has even followed him through the place names dotting the novel. But there's no documentation. In 1868, Hudson visited, according to Jorge Keen, the Keens' *estancia* called La Virgen de los Dolores in Soriano, Uruguay. The name changed in the novel to Estancia de la Virgen y de los Desamparados in Paysandú, but that's all. Nowhere has Hudson told us why he chose the buffer state Uruguay to write his romance, except maybe that it was off the literary map. In his day the Uruguayans were known as *Orientales* (Easterners) and so the suppressed chapter one was titled 'A Flight to the East'.

Here we read that the English Invasions of 1807 'burst like a terrible thunderstorm'. The original title – 'that England Lost' – obviously emerges from the historical proximity of the failed English invasions of the River Plate, blotted from English history. Here I got a clue as to how to interpret the protagonist Richard Lamb, for he was the son of a sheep farmer from the pampas who rented land, much like the Hudsons, and was married. In late 1870s London, Hudson relived the adventures and the freedom that he had lost.

There's more. The whole novel is a translation. All the people met and the stories told are in Spanish, which Hudson translated literarily. Lamb speaks Spanish so well that he may be a foreigner, 'yet one of us, a pure Oriental'. He rides, drinks *mate* and sits on cow skulls. His first night is an encounter with the bed bug the *vinchuca* (which possibly bit Darwin and caused him a lifetime's illness). He translates a song by Epifanio Claro. He tries to introduce milk-drinking, and associates with ten Englishmen, who fox hunt. But he distances himself from them so he can boast that it is not 'every wanderer from England who can make himself familiar with home habits'. When chased by a bull, he adopts a 'gaucho habit', feigning death. He is constantly scared of the dogs as he approaches a *rancho*. After kissing Transita, he defines himself as passionate, like the weather, brave to rashness, abhorring restraint, and loving women. He was born in the wrong country.

Set around 1858, the novel mutates from the original longing for a third English invasion to one of keeping the land free from Anglo-Saxon interference. At one point Lamb stops off with fellow countryman John Carrickfergus and his native wife Candelaria. He learns about an ideal education, avoiding books ('Read! What! . . . No, no my friend, never read'). Carrickfergus rejected his puritanical background, married, had six healthy children, 'dirty as they like to be', and taught them nothing. 'All we think about in the old country are books, cleanliness, clothes; what's good for soul, brain, stomach; and we make 'em miserable. Liberty for everyone . . .' The chapter echoes Hudson's own happy family of six offspring, and his education in freedom.

And then there are encounters with at least fifteen women. Lucio Costa Herrera, a naive Uruguayan reader, found Hudson's portraits of *criollo* women convincing. Lamb was a lady-killer, with good, tough looks so that Miguel de Unamuno could cry out: 'These women are the sole reason why the book lives.' But I found them unappealing stereotypes, not rounded characters. Borges thought that Hudson was not a first-rate novelist and 'romanticized the Uruguayan back country hopelessly'. I would add that he romanticized his women hopelessly.

Paquita by Keith Henderson.

But Borges also said *The Purple Land* was one of the few happy books ever written. He said that the most memorable phrase he'd ever read was from Hudson's Lamb. He does not give the phrase in full but it's, 'I have not read many books of philosophy, because when I tried to be a philosopher "happiness was always breaking in," as someone says . . .' So much for Borges's most memorable phrase in all literature. It seems that Hudson was quoting someone else.

Hudson deliberately hid the author of this quotation for he was not a literary show-off. Borges tells us it was Boswell, but improved upon almost to 'la perfección' by Hudson. How did Hudson improve Boswell? Well, Boswell was quoting Dr Johnson: 'when asked about the progress of his philosophical studies, confessed that he had given up "because cheerfulness kept breaking in"'. Hudson changed 'cheerfulness' to 'happiness'.

Chapter 6

Animals and Birds of the Pampas

Thousands of roaming and feeding cattle on the pampas is a strange sight – it's to do with the scale of it compared to English fields where individual cows are more appreciated, even have names. But even more distinctive was the mass lowing of the pampas.

In Devon, near Ottery St Mary, out rambling with his knapsack, Hudson came across a cowman driving pure Devons. Hudson admitted that cattle lowing 'is more to me than any other natural sound – the melody of birds, the springing and dying gales of the pines, the wash of waves on the long shingled beach'. But in England the cattle were silent compared, he told this Devonian cowman, to the cows 'of a distant country where I had lived' (he just couldn't mention Argentina).

Hudson called this mass lowing 'cow-music'. I have heard over 800 cattle lowing at the same time on the pampas. That was cow-music indeed. This cow-music is not a heavenly choir. It's more to do with incorporating all sound, even cacophony, as part of nature and not picking on what's sweet, like nightingales singing. Many birds do not have wondrous voices, in fact are inharmonious, but the sounds are still beautiful to Hudson.

Another kind of music comes from the cow-birds. Hudson especially adored their concert in the wood behind Las Acacias. This deep purple-black cow-bird, like a starling, is a parasitical cuckoo bird and lives in flocks. In his mind's eye, in London exile, he saw them feeding like a 'huge black carpet spread out on the sward'. These birds would sit in their thousands on branches and sing in a 'continuous torrent of song', a hollow guttural sound followed by loud clear ringing notes, a kind of mad exuberance 'flinging out their notes at random, as if mad with joy', he wrote.

I once stood by a wood at Las Tres Marías as the sun set and cow-birds packed the trees at the edge, singing their outrageous chorus as if competing with the sun itself. I found this hubbub of song frantic, neurotic, modern, atonal . . . and I ran out of adjectives so wrote a poem published in Spanish. These cow-birds outdoing each other in frenetic song reminded me of why Hudson wanted to become an ornithologist.

English critic H. J. Massingham compared Hudson to a trapped eagle, 'noble, melancholy, remote'. The portrait painter William Rothenstein compared him to a caged eagle in the zoo, a desolate prisoner, wings unused and drooping. Morley Roberts compared him to a condor in a cage. All three – and more – saw him as an exiled king of birds, trapped in a cage, in a city, in a foreign country. He did look like a bird of prey in older age, with his beaky nose and large shoulders like folded wings. But Luis Franco, an Argentine biographer, saw him as a pampas ostrich, a *ñandú* (from the Guaraní of Paraguay) or common rhea (*Rhea Americana*). Here's another bird not famed for song.

Johann Moritz Rugendas, *Boleando avestruces*, c1845.

In the closing entry to *Birds of La Plata*, Hudson pointed out how peculiarly adapted this fast-running bird is to the level pampas. He watched rhea hunting, with *boleadoras* or round stones tied with long leather thongs that were thrown round the legs of these birds to topple them, and observed their 'majesty and quaint grace'. They can be domesticated. The young males make love by twirling their necks around each other and then pecking hard. Eggs are laid at random in natural depressions. When they hatch, the parent rhea is dangerous to approach 'as the bird with neck stretched out horizontally and outspread wings charges suddenly, making so huge and grotesque a figure that the tamest horse becomes ungovernable with terror'.

In *The Naturalist in La Plata*, Hudson pointed out the rhea's unique, puzzling habit of running when hunted with one wing raised vertically like a great sail. He added that there were few more fascinating sights in nature than that of the old, black-necked cock rhea, standing with agitated wings, calling its scattered hens 'with hollow boomings and long mysterious suspirations, as if a wind blowing high up in the void sky had found a voice'. I see Hudson imitating their loud snorting sounds (Hudson would often speak back to birds), when the young birds would be fooled and rush to him for protection.

In *Far Away and Long Ago* the rhea is the 'greatest and most unbird-like bird of our continent'. I stress that use of 'our', which he repeats, saying 'our rhea'. Once when he was eight, on his pony, he was asked to keep an eye on sheep. He bumped into some rheas, their grey plumage so matching cardoon bushes that he missed them. He tried to catch one, but they were far too cunning for the young boy. Hudson recalled a saying in Spanish: '*El avestruz es el más gaucho de los animales*', meaning as cunning as any gaucho.

He lamented their almost complete extermination, but he was wrong. I've seen them countless times. *Ñandúes* continue to feed in flocks and melt into the grass with their greyness, heads on long necks, always watching us. Then they rush off, faster than a horse. They are seen as a pest as they flatten wire fences.

They leave their yellowish eggs around. Once a traditional *peón*, in his dark beret, turned up with a large egg he'd found for me on his horseback rounds of the cattle *estancia*. You can eat them by cooking them in their large shell, as Perito Moreno, the young Argentine explorer, did with the smaller *Rhea darwini* in Patagonia. I pricked a hole and let the yolk run out and preserved the eggshell. Then we scrambled this rhea egg, equivalent to twenty hen eggs, and ate it.

However, Hudson cut his teeth as a budding ornithologist on a pampas woodpecker. When Hudson wrote his early book on *British*

Pampas woodpecker, or Carpintero campestre, *Colaptes campestroides*.

Birds, the green woodpecker, this 'beautiful woodland' bird, with its tell-tale tapping on wood, but especially its call, was evoked as pure freedom:

> It is a clear, piercing sound, so loud and sudden that it star-
> tles you, full of wild liberty and gladness; and when I listen
> for it and fail to hear it in park or forest, I feel that I have
> missed a sound for which no other bird cry or melody can
> compensate me.

So much for the yaffle's raw cry; not a pretty songster.

However much Hudson identified with the English wood-pecker's wild cry of freedom, it was the pampas woodpecker or *carpintero* that haunted him. This clash over the status of a woodpecker became a parable about humiliation that began as a minor triumph.

The story is worth repeating. Gordon Wasson and Edwin Way Teal in the *Times Literary Supplement* of 1947 were the first to resolve Hudson's blank years of 1860 to 1880. These researchers found correspondence between Hudson and the Smithsonian Institute that established an itinerary. In 1865 the US Consul in Buenos Aires, Hinton Rowan Helper, writing to Dr Spencer Baird, eminent American ornithologist, introduced Hudson as an 'amateur ornithologist' and asked if he could be employed and paid.

In 1866, aged twenty-five, Hudson wrote his first letter to Dr Baird from his Quilmes shack. He had printed handwriting, far more legible than his later letters. He said he was inexperienced in skinning birds and would send snails and, especially, ostrich eggs. All the skinned birds sent had been shot within eight leagues of Buenos Aires. However, he was 'not a person of means'. Wasson noted Hudson's unconventional spelling. He had found six letters, but sixteen are now held at the Zoological Society's library.

Hudson began his writing career with this 1865 letter, imitating Gilbert White's epistolary natural history. He sent over 500 bird skins to the Smithsonian and was paid with $60 gold. But he was always hard up. Indeed, he told Baird on 15 March 1868 that he had 'no leasure [*sic*] to collect'. On 2 November 1868, he earned so little it 'scarcely pays my expenses' and added, 'I shall probably soon be obliged to give it up.' He managed to sell some skins to collectors in Buenos Aires, and sent Baird 205 skins after four months at Ensenada de Barragán. But he had been on military service and working as a gaucho most of the time. He'd first turned to the United States for help, maybe feeling more American than English.

Dr Hermann Burmeister.

In October 1865, before writing to Dr Baird, the twenty-four year-old Hudson had made contact with Dr Hermann Burmeister in Buenos Aires, who had appointed him as a collector to explore Andean passes, but was then told the Argentine government couldn't pay him. Hermann Burmeister (1807–92), a German naturalist recommended to the Argentine government by Humboldt, could be called the first professional scientist in Argentina. He edited a journal, ran the Museum of Natural Sciences from 1862, and published studies in the wake of his 1843 *History of Creation*, which had made his reputation in Europe. According to Dr Philip Sclater of the Zoological Society in London, he was 'as well known in Europe as in Buenos Ayres'.

He'd arrived in Argentina in 1857 and by 1861 had settled. He was the first to systematize Argentina's natural history, based on his travels, in several volumes written in German and published between 1876 and 1886. But his fate was sealed, because he opposed

Darwin's historical explanation of species division, and publically fought the local Darwinist Florentino Ameghino. However, Burmeister had fallen in love with Argentina and didn't want to return to Europe. He was a famously grumpy and authoritarian figure, complaining about never being paid and that he'd had to farm on the side to earn. He fought *criollo* trickery, like when his horses were stolen by night and sold back to him next day.

Hudson could be called Burmeister's 'disciple', as he set him on his scientific path. Through Burmeister he acquired a taxidermist's manual and bought Félix de Azara's *Pájaros de Paraguay y del Río de La Plata*, 1802, which he quoted from for the rest of his life (Darwin also admired Azara).

By luck, Dr Baird of the Smithsonian then recommended Hudson to Dr Philip Sclater of the London Zoological Society, who subsequently read Hudson's letters at the society's meetings. London's Zoological Society was at the centre of the scientific

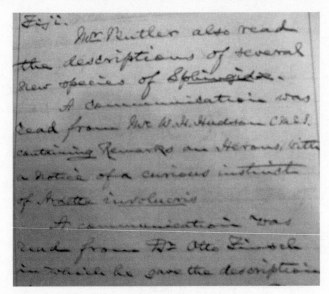

About W. H. Hudson, 16th November 1875,
Zoological Society Minutes.

world, so it was a good move. Even Dr Hermann Burmeister sent new specimens there.

Hudson's letters, held in two envelopes at the Zoological Society's library, are written on lined paper, clearly best copy as there are no crossing outs, but the ink has faded. They were edited, copied out and then printed, so Hudson's first letter of 1868 was received in London in 1869, five years before leaving for England. That same 1868, Philip Sclater and Osbert Salvin, eminent English ornithologists, published a list of the 265 bird-skins Hudson had collected at Conchitas in three collections totalling 145 species. They hoped that 'Mr Hudson will continue his collection in this interesting locality'.

In 1870, Hudson wrote complaining that only four out of ten letters had been acknowledged by Sclater. Maybe, he wrote, his letters were being *expropriated* by the Post Office'. Letters took ages to arrive. One he sent on 19 May 1870 reached London on

Dr Philip Sclater.

15 November. But not all was frustration for him. The same year, 1870, Sclater made him a Corresponding Member of Zoological Society (CMZS), naming two Patagonian bird species after Hudson: *Cranioleuca hudsoni* (renamed *Aesthenes hudsoni*) and *Cnipolegus hudsoni* (now *Phaeotriccus hudsoni*). In his *Birds of La Plata*, Hudson noted the latter's fantastical dance as it whirled around its branch perch, uttering sharp clicking noises, adding 'all the movements of the bird are eccentric to a degree'. Nothing was known of its habits. That summed up Hudson's privilege in being on the spot in Patagonia, for little was known of that bird life, unlike British birds which already had a long, ornithological history.

He would always proudly add CMZS after his name. It was an honour awarded to members who didn't reside in England. Later, in 1876 and resident in England, he would become a Fellow of the Zoological Society.

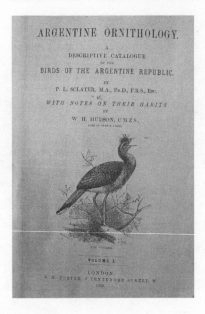

The librarian at the Zoological Society brought me the leather-bound book where I found Hudson's Form of Recommendation for a Fellow, number 9303. He had his name back to front as Henry William Hudson, and he'd been recommended by Dr Sclater, D. Richardson and Robert Hudson (a namesake but not related). This society's acronyms after his name were Hudson's badge of scientific respectability as he didn't have a university degree. At their meetings, he rubbed shoulders with the world's leading scientists. For example, Hudson attended one meeting on 4 January 1876 where the communication before his own was read out was by Professor T. H. Huxley, Darwin's 'bulldog'. Darwin himself had been elected a Fellow in 1831.

Dr Philip Lutley Sclater, a barrister and zoologist, educated at Winchester and Christ Church, Cambridge, was the right man to help Hudson before he left Argentina and later was the first person he contacted on arrival in England. He was founder and editor of the magazine *Ibis* and secretary to the Zoological Society in 1860–1902. He twice travelled to the United States, corresponded with everybody in the scientific world (he was a gifted linguist) and by 1896 had published over 1,300 scientific papers. He dedicated himself to encouraging others. His office at 11 Hanover Square was a focal meeting place and he was a man of 'great power'. Hudson had the luck to know this influential man of science.

Crucial for Hudson was their collaboration in the two-volumed *Argentine Ornithology*, 1888 and 1889, giving the books a solid scientific gloss. Hudson wrote personal notes about 434 species, while Sclater busied himself with the scientific nomenclature. Two hundred copies were printed. A review in *Nature* claimed that Hudson and Sclater 'have combined their forces to produce one of the best books ever written on South American ornithology'. Hudson was 'one of the best living observers of the habits of birds in the field'. In Sclater's words, he was 'my fellow-author, though English in name', but 'an Argentine citizen by birth'. He had collected in the pampas, the woods and marshes along La Plata, the sierras from Cape Corrientes to Azul and Tapalqué and Río Negro in 1871.

Only a few personal anecdotes about Sclater and Hudson have survived. One has Hudson telling Morley Roberts: 'Old Sclater used to say "Let a man be as humorous and witty as he likes but he must keep all that out of a scientific paper"'. It's still the case. Another anecdote concerns something that happened between the two men, possibly related to social class and education, for Hudson told the poet William Canton that Sclater and himself were 'two men who have not one thought or taste in common'. But this note was contradicted by Hudson's unpublished letter to Sclater at the Zoological library, dated 1882, where he thanked the secretary for encouraging him and for his immeasurable kindnesses. In another unpublished but personal letter, he posted him his book *Nature in Downland*, 1901.

Maybe he was being polite. When Hudson reissued the early work as *Birds of La Plata* in 1920, he simply cut out Sclater's contribution. By then, he was long dead.

The early dealings with Sclater set the template for Hudson the naturalist. He read Hudson's letter dated 14 December 1869 aloud at a meeting of the Zoological Society on the 10 February 1870, and then published an edited version. It's fascinating that from this first publication, Hudson had found his voice in acute, subjective observations of quirks in nature in the form of anecdotes. He is very much in the picture, not an emotionally detached, objective naturalist. He didn't sound like a scientist. Here is the man himself (he later pasted this letter into *Birds of La Plata* in 1920):

The first bird of this species I shot (Chestnut-shouldered Hang-Nest), was but slightly wounded in the wing, and fell into a stream; to my great surprise it began singing as it floated about on the surface of the water, and even when I had taken it out, continued to sing, at intervals in my hand. I subsequently found a nest of this bird, it was about seven inches deep, composed entirely of lichens, curiosly [*sic*] wove together and suspended from the twigs of a low tree.

It is from his second letter to Sclater, dated 22 December 1969, that Hudson began his attack over the Pampas woodpecker 'of which Mr. Darwin has so unfortunately said – "It is a woodpecker which never climbs a tree"!' The next letter, of 28 January 1870, expanded this daring critique. Hudson was twenty-nine years old.

Darwin by George Richmond, 1840.

His attack shocked Sclater, who published it, but toned down Hudson's cheeky aspersions. Remember, he was an unknown bird lover from Argentina.

Hudson announced that of the four woodpeckers in his country, it's the *carpintero* he wanted to bring to his English readers' attention because of 'the erroneous account of it in Darwin's great work'. Hudson was so sure of himself that he praised Darwin for being a close observer, but undermined it by complaining that he didn't stay long enough in Hudson's parish of Quilmes in his 'rapid ride across the pampas'. He was perhaps deliberately evoking the title of Sir Francis Bond Head's 1826 *Rapid Journey Across the Pampas*. But what did Hudson mean by his jibe of a rapid ride?

On 3 August 1833, the *Beagle* anchored at the mouth of the Río Negro, with El Carmen eighteen miles upriver. Darwin decided to ride overland to rejoin the *Beagle* at Bahía Blanca. Darwin, with an

Juan Manuel Blanes, *Los Boleadores*, c1876.

English interpreter and five gauchos, rode to the Río Colorado, where he met and interviewed General Rosas, who was harrying the nomadic Indians at the time.

On the way, Darwin had glimpsed the famous Walleechu tree that Cunninghame Graham properly called the Gualichú tree in his sketch of that title. This thorny, solitary tree was loaded with suspended gifts from Indians to their deity, from cigars to bits of rag, surrounded by the bleached bones of sacrificed horses.

Darwin evoked his first night out under the stars: 'The death-like stillness of the plain, the dogs keeping watch, the gipsy-group of Gauchos making their beds round the fire, have left in my mind a strongly-marked picture of that first night, which will never be forgotten'. As he rode through Indian territory, Darwin lamented the extermination of the Indians, especially the women because they bred. He noted in his diary: 'Who would believe in this age that such atrocities could be committed in a Christian civilized country?'

He remained in Bahía Blanca, waiting for his ship, then decided to ride the 400 miles to Buenos Aires. He took one gaucho as guide, was later joined by a small military escort and galloped off towards the rocky outcrop of ancient granite called Sierra de la Ventana. He became the first foreigner to climb it, ate armadillo in its shell and the white meat of a puma, saw thirteen deer killed by giant hailstones and noted how gauchos only ate beef. He didn't eat ostrich egg, though. He reached Monte's rich green plains on the outskirts of Buenos Aires, with its *estancias* and their solitary ombú tree. Finally, on 20 September, he entered Buenos Aires.

A rapid ride? Perhaps, but from 11 August to 20 September 1833 – forty-one days – Darwin had lived as a gaucho, head on saddle at night, eating what could be shot or caught with a *boleadora*, and riding. This excursion into danger does not match Hudson's jibe of 'rapid ride across the pampas'.

In this 1870 letter Hudson also attacked Darwin's theory of natural selection or speciation: 'Certainly if he had truly known the habits of the bird, he would not have attempted to adduce from it an argument in favour of his theory of the Origin of Species'. He added insult to injury by saying that one 'deviation from the truth' (about woodpeckers) suggests many further errors, thus hoping to arm or even join 'opponents' of his book like Burmeister, his mentor. He mocked the 'easy' way natural selection alters an animal's habits. He asked how could a bear mutate into a whale. There was no evidence for intermediate species. His pampas woodpecker 'lends no favor to this part of the argument'. From Quilmes, Hudson denied that a woodpecker, so perfectly adapted to its tree and niche, could live in the pampas where no trees grow. As he was better acquainted with the woodpecker and its habitat, no doubt Darwin 'purposely wrested the truths of Nature to prove his theory'. He found similar mistakes in his *Beagle* book about 'this' country.

Of course, he wrote, woodpeckers frequent trees, orchards, willows, in fringes of woods along the River Plate shore and forests. Even the ombú is home to the bird. As we know, there were twenty-five ombúes by his hut. His letter bursts into life:

> This tree (the ombú) attains a considerable size; there is one within fifty paces of the room I am writing in, that has a trunk, measuring, three feet above the ground, thirty feet in circumference. This very tree was, for many years, a breeding place for several Carpinteros . . .

He returned to the attack on Darwin as his long letter closed. It's a David and Goliath moment: 'It is not only the altogether erroneous account of this bird's habits that makes Darwin's mention of it peculiarly unfortunate for him, but also because it is rather an argument against the truth of his hypothesis.'

In a postscript that Sclater didn't publish, Hudson was forced to admit he couldn't give Latin names as his authority on Argentine birds, Burmeister, had not been translated from the German and Burmeister himself was gravely ill.

Darwin's courteous reply to Hudson's criticism was published in the *Zoological Proceedings*, part three, in 1870. His son Francis Darwin had been surprised that his father 'departed from his practice' in answering a critic like Hudson. Why did Darwin react this one time to an obscure naturalist from La Plata? Could it have awoken pleasant memories of his *Beagle* trip or could it be that Hudson had touched a sore spot?

When Hudson read Darwin's reply is not recorded. It's certain that Dr Sclater sent the journal to Buenos Aires, because Hudson acknowledged copies of the *Proceedings*. In this letter, Darwin named the woodpecker in Latin as *Colaptes campestris*. He insisted he

Darwin's reply, *Zoological Proceedings*, 1870.

repeatedly saw this woodpecker many miles from trees. He shot some and studied their muddy beaks. He consulted Félix de Azara, an 'accurate observer', who called this a *pico del campestre*, a woodpecker of the plains, as it never visited trees and fed in the open, was in fact a slightly modified species, building nests in holes on the banks of streams or in old mud walls. But Darwin was forced to admit Hudson was 'perfectly accurate' and that he, Darwin, should not have used the word *never* about a woodpecker climbing trees. He ended with an ominous accusation: 'I should be loath to think that there are many naturalists who, without any evidence, would accuse a fellow worker of telling a deliberate falsehood to prove his theory.'

In that year 1870, before arriving in England, Hudson still dreamt of becoming a professional field naturalist – after all, he had been paid for 500 bird skins, was being published by the Zoological Society in London and had been honoured with a CMZS after his name. It was the most exciting time of his life. He sailed to England in 1874 with that ambition in mind. He could have made it as a 'fellow worker', had he believed in Darwinian evolution by natural selection. I know that by Darwin's death in 1881, six years after landing at Southampton, Hudson's dream was over. In a census for 1881, he'd defined himself as a writer and journalist.

But Darwin went further. In the sixth, 1872 edition of his *On the Origin of Species* he found the incident with Hudson faintly amusing. I recall my shock on reading the book at seeing a 'Mr Hudson' appear, without any further identification. That must be W. H. Hudson, I exclaimed. Darwin had modified his generalization about woodpeckers *never* climbing trees, according to Hudson's 1870 letter. But after praising Hudson as an 'excellent observer', he damned him as 'a strong disbeliever in evolution'.

Hudson had reached England two years after entering the most influential book of the nineteenth century. By that date, there would be no room in scientific circles for a 'disbeliever' in evolution. When Hudson turned up at Dr Sclater's offices in 1874, he did not know his fate, but years later when he wrote his entry on this woodpecker for his *Birds of La Plata*, 1920, he finally had to agree with Darwin, who had never been wrong: 'They also seek their food more on the ground than on trees in some cases not at all on trees, and they also breed oftener in holes in banks or cliffs than in the trunks of trees.' But not a word about the dire consequence of his 1870 letter and having been tarred as a *disbeliever in evolution*.

What Hudson could not comprehend from Quilmes in 1870 was that the woodpecker had been a secret test case for Darwin. He hadn't a clue that in 1860 a 'very hostile' review – Darwin's own words – of his *On the Origin of Species* appeared. Andrew Murray denied Darwin's idea of a treeless woodpecker and said it shouldn't be called a woodpecker. Darwin modified the 1862 edition to placate Murray's critique. He did the same three years later with Hudson in 1872, and again over a woodpecker.

Furthermore, Hudson could not know how deeply Darwin had admired clergyman William Paley's creationist *Natural Theology: or, Evidences of the Existence and Attributes of the Deity*, 1802, where the woodpecker featured as a bird exquisitely adapted to its niche. In a letter to J. Lubbock in 1859, Darwin confessed he knew Paley's book almost by heart. A woodpecker without a tree, all its intricate mechanisms rendered useless, was a species re-adapting, a 'sport'. Sir Gavin de Beer saw that 'Darwin's favourite example of adaptation' was the woodpecker. Paley's woodpecker perfectly adapted to a tree implied a God who had suddenly created a fixed world of intricate adaptations in six days:

> The peculiar conformation of the bill, and tongue, and claws of the woodpecker, determines that bird to search for his food amongst the insects lodged behind the bark, or in the wood, or decayed trees: but what would profit him, if there were no trees, no decayed trees, no insects lodged under the bark, or in the trunk?

Darwin knew Paley's message that we live in a harmonious, happy world where everything adapts to everything. But while travelling on the *Beagle* for five years, he discovered the La Plata woodpecker feeding on the ground, *without a tree*. Darwin had discovered a transitional form, an in-between species. In his notebook in 1837 Darwin knew he was on to something essential about life: 'The circumstances of ground woodpeckers – birds that cannot fly etc seems clearly to indicate those very changes which at first might be doubted were possible.' By 1838, Darwin knew that the woodpecker's claws, used to perch on tree trunks, were 'pointless'. He had discarded Paley's evidences.

At the end of his life, in his 1920 *Birds of La Plata*, Hudson feared that Argentine woodpeckers had become 'nearly extinct'. But he was wrong. These woodpeckers still live in the woods and trees around pampas houses. I've seen them on acacias and on the ground, and they even attack houses by drilling into brick and wooden shutters. In fact, they're a national pest. By 1920 Hudson knew Darwin had been right all along, only he refused to admit his hasty judgement of fifty years back that had decided his destiny.

Chapter 7

The Enigma of Arrival in England

What went through Hudson's mind during the years from his mother's death in 1859 to his final decision to take the boat to London in 1874 can only be guessed at through anecdotes he told and by letters written to the Smithsonian and the Zoological Society of London. For, apart from these letters, he wrote only notes while in Argentina. His writer's life began in England after 1874. That's why I have trawled for traces of the man in fiction written in England, like *The Purple Land* and *Ralph Herne*, as it explored elements of his Argentine past. I'd add that his fiction can be better read biographically as on its own terms it's derivative. But a lot is known about Hudson's journey over.

Hudson stepped onto English soil for the first time in his life on 3 May 1874. The Royal Mail steamship, the *Ebro*, had docked at Southampton and he would be thirty-three years old two months later. He would never return to his Argentine home. Later, he would once travel to Ireland but he never crossed the Channel. What did this aspiring Englishman look like? Only one photo exists of this

The *Ebro*, 1872.

youngish man back in Argentina. Most portraits were painted thirty
years later, when he had achieved some fame. Verbal descriptions
tend to describe old man Hudson.

From the outside he was a striking figure at six foot one, with a
disproportionately small head, hazel eyes, long dark hair, bushy eyebrows, a trimmed beard, a broken nose, deeply sun-tanned and
extremely fit from his outdoor years on the pampas. He never lost
his fitness, despite his illnesses. Years later in 1908 David Garnett
dressed this agile man of sixty-seven in a kind of uniform: rough
tweed jacket, with pockets in the tails, waistcoat and trousers to
match, shirts with starched stand-up collars, cuffs and black lace-up
boots. Hudson always resented this suit of armour. Like Charles
Waterton, he loved taking his boots off and walking barefoot. He
recalled how Benjamin Franklin got out of bed and sat down naked
to write. From his arrival in his thirties to his seventies, Hudson
maintained what Richard Curle called his 'lithe step of an athlete'.

The *Ebro* that brought him to England was built in 1872 with
three classes and 127 passengers. No passenger list survives. It was a
steamship, running on coal, but also had two sails and weighed 1,500
tons. The journey from Buenos Aires had lasted thirty-three days.
I assume he travelled as a gentleman, though poor, and had saved
from his meagre inheritance and from shepherding. He was not in

Earliest photo of Hudson in England,
c1876.

steerage, but probably was a second-class passenger, with students like Abel Pardo. However, he never harked back to this ship journey and, like his pioneering, American parents in Argentina, he'd set out to start a new life in England from when he first caught sight of the 'white cliffs and green downs of Wight and the Hampshire shore'.

All that is known of this sea trip is from a long diary-letter he wrote to his younger brother, Albert Merriam. I read the original, in neat handwriting, at the Hudson museum in Florencio Varela. We learn that the small vessel pitched so much that he could hardly write: 'Roll roll roll – morning, noon and night', but he'd found his sea-legs. The only sounds were the hiss of steam and the throb of the vessel. A few whales were spotted and one passenger caught yellow fever, which ensured that nobody could land at any of the ports of call, like St Vincent in the Cape Verde islands and Lisbon.

He became 'great' friends – his words to Cunninghame Graham – with the Argentine student named Abel Pardo, who later translated 'Paulino Viera's Confession' for *La Nación* in 1884. It would be

Opening page of his diary-letter home, 1874.

Hudson's first appearance in Spanish. Abel lent his name to the working title, 'Mr Abel', that later became *Green Mansions*, and to its protagonist. He lived in London for almost two years and later regularly visited, meeting Hudson, for example, in 1895 and 1898. He had grown up on an *estancia* and once posted poems about gauchos. Hudson called him, without naming him, in *The Land's End*, a 'very keen observer', coming from an old native family, and rising to a very high place in the government. Pardo once differentiated the drinking habits of the English, who drink to relax and talk, from the Argentinians, who are already relaxed and can chatter.

He was a prominent revolutionary in the Radical Party that set out to destroy the current system, as he wrote in a manifesto in 1891. He became a minister in 1893 in the government of the province of Buenos Aires. Hudson ignored politics and history, but as they dominated Pardo's life until his resignation in 1898, it's hard to believe they didn't talk about Argentine politics.

Only once did Hudson refer again to this sea journey, when, in *Nature in Downland*, he evoked the giant ocean swell.

Hudson did not rush off by steam train to London, but slowly accustomed himself to Dove's Hotel in Southampton, its surrounding countryside and the Isle of Wight. He was suffering from culture shock. His crucial question, 'How can I describe England?', began with an urban comparison. Southampton, with its population of 55,000, made Buenos Aires, a city of some 240,000, 'now seem to me a poor, filthy, rough, ugly, disagreeable city'. Nevertheless, he quickly learnt to adapt by buying his regulation umbrella and gloves. He tried bread and beer – a ploughman's lunch – at a pub. And he spotted birds: 'I heard thrushes, wrens and many others I have heard so much about'.

But nothing could prepare him for the thrill of late spring birdsong. He identified a cuckoo for the first time in his life. He'd spent years trying to imagine the sound of English songbirds. 'No person at a distance from England could have striven harder than I did', he wrote, 'by inquiring of those who knew and by reading ornithological works, to get a just idea of the songs of British birds'. But all his efforts were to no avail as words could not convey song, so that each bird he heard on arriving, like the cuckoo, came as a shock. Referring to the song of a blackcap, he wrote 'how in boyhood this same bird warbled to me in some lines of a poem I read; and how, long years afterwards, I first heard the real song – beautiful, but how unlike the song I had imagined!' Here was the key, I thought, to Hudson's nature writing. He had read about English birds, but seeing and hearing them for the first time was cathartic.

While on the Solent on board his steamship, passing the Isle of Wight, Hudson asked a sailor if he could point out Poet Laureate

Alfred Tennyson's house at Farringford. The sailor had never heard of Tennyson. So here was the second cultural shock: the English did not care much for culture. Or for geography outside the Empire, as he learned chatting on board with an aristocrat he mocked as Sir Lester Dedlack (distorting *Bleak House*'s Sir Leicester Dedlock), who had never heard of the River Plate or of Buenos Aires. Throughout his life, English insular ignorance about Argentina humiliated Hudson.

After five days roaming around his port of arrival, an Argentine passenger called Jimenes sent him a message that he had found an economical place to stay in London. From a passage in *Birds in London* we learn about his first week in a City hotel, when he had seen St Paul's Cathedral, the 'turbid, rushing Thames' and its bridges, and walked at random through 'miles of brick and mortar and innumerable smoking chimneys'. But he'd found solace amongst huge trees and a rookery in Kensington Gardens. On this wet, cold May morning in 1874, walking around the trees and grass, London traffic hubbub gave way to cawing rooks. Hudson called the later felling of some 700 of these giant trees with rookeries an 'unspeakable barbarity'. For the rest of his life in London, Kensington Gardens, with its sheep keeping the grass cut, would function as his retreat.

Hudson had surprisingly brought with him on the *Ebro* more than 200 artefacts from his year in Patagonia, which had survived the shipwreck of most of his collection (as well as essential books and his journals). He donated it in 1874 to the Pitt Rivers Museum in Oxford. A. H. Lane Fox catalogued and published this collection as flint and chert arrowheads and stone tools (but no skulls from the many found near Carmen de Patagonés). I reckon this gift was his token of belonging. He could have donated them to Francisco Pascasio 'Perito' Moreno, as he had with earlier finds in Quilmes, but kept them for Oxford.

Just before he died, he returned to write of his arrival in England in *A Hind in Richmond Park*, 1922 (Morley Roberts struggled to decipher his handwriting to complete the final chapter). England had an 'earth-born, thick, warm smell', which nobody he asked could identify. He called it the smell of England. Then, months later, walking down Oxford Street, near Tottenham Court Road, he located the smell issuing from a brewery. It was beer.

In his early days sometime in 1874–5, abandoned and alone, he met Emily Wingrave, who was his landlady at 11 Leinster Square, W2 (now a hotel). Emily had also lived at 16 Leinster Square. Today number 11, a tall, white Victorian building, is boarded up and run down, but it was where the Hudsons lived for eight years. Here is a fascinating event for there hadn't been anything in his life before this meeting that touched on his relationship with women. Out of sheer loneliness in the biggest city in the world, with a population of 4,776,661 and growing, he decided to marry this Englishwoman Emily.

Morley Roberts easily guessed why they married: 'a poor guest married a kindly hostess, then in some hazardous prosperity'. Emily was on the shelf, of an indeterminate age (probably forty) and dubious sexuality, but she sang beautifully as a soprano and was a music teacher. There's nothing about their courting, what they discussed, what kind of marriage they contemplated.

All we have is their marriage certificate stating they were married on 18 May 1876 at St Matthew's Church, St Petersburgh Place, Bayswater, near the Greek Orthodox church on Moscow Road and not far from a synagogue. The church stands tall with its spire. Its facade is begrimed with centuries of soot and needs a clean. Emily gave her age as thirty-six and Hudson also lied. He gave his rank as gentleman, his address as 40 St Luke's Road (also owned by the Wingrave family), his father as Daniel and his father's occupation as farmer (but not that he was dead). Emily lived at 16 Leinster Square and her father John Hamner Wingrave was Accountant General in HM Civil Service, and he signed in front of James Hunter, the vicar, as one of two witnesses (the other was Louise Hamner Basset, possibly an aunt or married sister). Hudson's marrying this Englishwoman when he was at his most vulnerable and desperate was his first step towards assimilating with the English.

As for his dreams of becoming a scientist in England, Brian Tippett acutely remarked that it must have been heartening for Hudson, with so few to discuss birds with, to find himself a member of the distant scientific community of the Zoological Society before he even left Argentina. Hudson was adamant: 'in all the years of my life in the pampas did I ever have the happiness to meet with anyone to share my interest in the wild bird life of the country I was born in'. The Buenos Aires he left behind was a city expanding with European immigration from 90,000 in 1855 to 286,000 in 1880, but parochial compared to London. He had tried to correct Darwin and had two bird species named after him, so, like a Balzac character, set out to conquer this immense and inscrutable city.

The first person Hudson visited, naturally as he'd been corresponding with him, was Dr Philip Sclater at the Zoological Society at 11 Hanover Square (it moved to 3 Hanover Square in 1883 and shared the building with the Society for the Protection of Birds). But despite Sclater's obtaining £40 from the Royal Society for Hudson to co-author a book on British birds and supporting him in becoming naturalized, the relationship didn't prosper. In fact, it dawned on Hudson early on that he wouldn't make it as an ornithologist and meeting Sclater was the proof. In *Adventures among Birds*, 1913, he finally admitted that his dream of making a name for himself pursuing wild birds, 'my one desire', was 'never in my power'.

There are glimpses of the jobs he undertook in his first years. One was with the ornithologist, embalmer and artist John Gould (who identified the birds Charles Darwin brought back from his *Beagle* trip) in Bedford Square. Gould was in his early seventies. Hudson especially resented his famous collection of stuffed hummingbirds (some 5,000 of them in twenty-four cases made by Gould himself, in the Natural History Museum), because he had seen countless such beauties alive on the pampas. Hudson called Gould's hummingbirds 'pellets of dead feathers, which had long ceased to sparkle and shine'. In *The Naturalist in La Plata*, 1892, he evoked the speed of this flower-sucking bird as it flits and vanishes, a picture 'of airy grace and loveliness that baffles description'. But this unbelievable beauty fades when the bird is dead.

Stuffed birds were 'revolting to me', he later wrote, and cited St George Mivart, an anti-Darwinian scientist, on there being no such thing as a dead bird. The life is the bird. No art or taxidermy can resurrect the swift motion of wings that blurs the hummingbird. His target was Gould, whose colossal monograph on British birds only represented dead birds. Gould's nearly 3,000 bird drawings dominated his field, and he was ruthless, as Hudson knew to his own cost.

Edward Lear, who also worked for Gould, found him a 'harsh and violent man'.

Another poorly paid job was as secretary to an indebted archaeologist called Chester Waters. Hudson had to research genealogies for American tourists, but in fact also had to sneak food into Waters's house, as the latter avoided debtors. In a letter Hudson referred to Waters gossiping about a distinguished writer he knew who wrote one sentence a day.

In these early London days, Hudson read as much as he could about England's natural history. Whenever possible he escaped from the grime and soot into the warmth of the blue-domed British Library. He took out a reader's ticket in 1875, giving his address as 40 St Luke's Road, WII, where he and Emily would eventually live. His occupation still was 'naturalist'. There's a record of taking out another reader's ticket in 1883 while living at 11 Leinster Square. Over this first decade in lonely London, he wrote *The Purple Land that England Lost*, and published several poems. Like the impoverished Karl Marx, working on the second and third volumes of *Das Kapital*, Hudson was building up his mental library. He had no other job. I like to think of Marx and Hudson researching near each other at their desks. They had nothing in common, and Hudson wouldn't even try to read him.

I've no doubt that his first London friend was his wife and rambling companion Emily Wingrave. She was very short – reaching Hudson's elbows, said Ford Madox Ford, who compared her to a vivacious hummingbird at his stooping side. She had a mane of red hair, which Hudson appreciated, and was once a professional opera singer, which again Hudson valued, as hearing topped his list of senses. He told poet Wilfrid Blunt in 1921, at the end of his life, that he'd married because 'her voice moved me as no singing voice has ever done before'. She had sung with Adelina Patti (whom Verdi called the greatest soprano he'd known) and Sims Reeves, until she lost her singing voice. Unfortunately, after some research, I could neither confirm nor deny that last comment, as her name has never appeared in any programme or document I can trace.

Hudson wrote to his niece Maggie Hudson on 21 May 1921 that, after her death, he missed his wife 'terribly' and that she was 'the nearest to me of all beings on earth'. She and Hudson wrote letters to each other every day they were separated. He closed one of the few surviving letters that I read in the RSPB library with 'Goodbye, darling, with love & *usual* kisses'. She also read much fiction and was quick-witted, even quick-tempered. Hudson often commented that she was the first to read the novels sent to him. She once contrasted her husband with Richard Jefferies, showing critical acumen, before it became a commonplace. I wonder if she was involved in Hudson's writing, by listening and correcting his English in the early years.

Alicia Jurado concluded that she was a very dedicated and loyal wife. However, there is something amiss about her concealing her age and not having been married before. She did cultivate the illusion that she was 'ageless', as a friend observed. Could Emily have been a secret lesbian? In Hudson's strange novel *Fan*, Miss Mary Starbow, based on Emily, has lesbian tendencies. It might explain Hudson's sexual reserve (narrating through a young girl, Fan) and Emily's odd appeal, with her crush on Fan. That they were intimate 'friends', rather than sexual partners, perhaps is true. But I tend to agree with Alicia Jurado over *Fan* that to modern eyes they 'all seem lesbians', which would have shocked Hudson.

Comments on Hudson's wife were not flattering. His niece Laura wrote home from her honeymoon in 1910 to say she found Emily 'small, white and ugly', but got to like her during the five-day visit. Violet Hunt, who coquettishly befriended old man Hudson, found her squat and dwarfish, with large, ugly, ill-kept hands, but she was

a snob, viperish according to David Garnett, and an indiscreet chatterbox, and Emily was then old and infirm.

Emily's birthdate is a mystery. In the 1861 Census she was born
'about' 1836 in Lambeth, Surrey. Her father was a retired Chief
Accountant at the Inland Revenue (Hudson claimed they lived at
Somerset House). In the 1871 Census she was thirty-five and her
father had died. What month she was actually born escapes the
documents, but her birth was registered in the Oct–Dec 1837 slot.
So she was around forty when she married, but clearly lied in the
marriage certificate, claiming to be thirty-six. No biographers have
established her real age, and accounts vary from being twelve to
twenty years older than Hudson. She concealed her age to the end.
On her tombstone only the date of her death was recorded.

I scoured the internet for documented threads of the Hudsons'
life together. In the 1881 Census they lived at 11 Leinster Square
(where they lived from 1876–1884 and before that briefly at
16 Leinster Square). They also lived at 5 Myrtle Terrace in
Ravenscourt Park where Hudson once confessed that they were so
poor they lived off a tin of cocoa and milk for a week. Hudson
enjoyed overlooking the private park's wild green expanse. He also
lived at further boarding houses. I found he'd given the addresses
22 Craven Terrace and 17 Cleveland Square to the Zoological
Society. He also once lived near Kew. In the 1881 census Emily was
a 'teacher of singing', with a bureaucrat scribbling 'music' above
her name.

Years later, Sir William Rothenstein was surprised to learn that
Hudson was married. He wrote in his memoirs that 'one day he
[Hudson] spoke of his wife. "Married!" said my wife, "and you never
told us. How long have you been married?" "As long as I can remember," was Hudson's answer, the gloomiest verdict on married life I
have ever heard.'

Though no date is given, Hudson was by then an old man and Emily an invalid. To ensure their intimacy remained hidden, Hudson burnt most of their correspondence. I hesitate to speculate about their sex life or whether it was a *mariage blanc*. But she was an opera singer, used to male suitors, and he was a lithe and muscular man from the pampas.

On the same 1881 Census, I found that he too concealed the year of his birth (1841), putting 'abt' 1842 in 'Buenos Ayres'. However, his occupation was 'Writer for Periodicals (Journalist)'. So clearly, his scientific ambitions had collapsed before this census. By the 1891 Census he was still unsure of his birthdate, born 'abt' 1843 and had become just 'author'. By the 1901 Census, Hudson was naturalized British, was fifty-nine and Emily a year older at sixty. Even the 1911 Census wrongly states that Hudson was married in 1875 and was born in Ayr, Scotland (that jolted me), but someone had obviously mistranscribed Buenos Ayres. He still lived at 40 St Luke's Road and still estimated his birth wrongly at 'abt' 1842, but his job was 'author and journalist'. Is that what Hudson meant by 'Perhaps I may say that my life ended when I left South America', that it was by leaving the vivid experiences of the pampas that he turned into a writer? His friend the poet Edward Thomas thought this was the essential clue: 'Mr Hudson began by doing an eccentric thing for an English naturalist. He was born in South America.'

For 'writer' he had become almost as soon as he'd arrived. In 1875 he published an essay, 'Wanted, a Lullaby', in the women's magazine *Cassell's Family Magazine* (never republished). It raises fascinating questions of identity. Hudson pretended to be a woman called Maud Merryweather, who had a baby, and had lived 'much of my life' on the *west* coast of South America, deliberately avoiding naming Argentina. She had learned from the Tehuelche Indians how to sooth a babe to sleep with a lullaby, which 'she' translated from the

Spanish lullaby 'A-ro-ró mi niño'. Like Hudson himself, Maud loathed 'the ultra-refinement of our social life'. She hints on how to sing English babes to sleep, adding that the notes 'were written in a distant land, where there is another nature'. Hudson/Maud refers to the song 'Far, far away' that lodged in his memory and years later would title his memoirs. Why did Hudson pretend to be a woman? Was he keeping his name for serious natural history?

I'm still puzzled as to why Hudson was as reluctant to admit his age as Emily, but offered no doubt about being an Argentine born of American parents, a 'child of an alien and heretic race'. However, he was also an alien in Britain until awarded his Naturalisation Certificate on 5 June 1900, where he was confirmed as 'author and journalist', as on the censuses.

By marrying Emily Wingrave, he had a home. Emily finally inherited a mortgaged 40 St Luke's Road (also known as Tower House) from her sister in 1886 and that became home. Bizarrely, Hudson had lived there on his own, under another landlady, in 1875–6 as we saw. Near Paddington station, it was, in Ford Madox Ford's words, 'a fantastically gloomy house in the most sooty neighbourhood of London'. Rothenstein also found Tower House to be large and dreary. Hudson remained so poor that he refused to hobnob with the rich, making a few exceptions like Mrs Bontine, Cunninghame Graham's mother. She ran a literary salon at 39 Chester Square, and outlived Hudson, attaining 101 years old. The Hudsons occupied two floors and let off the rest to pay the mortgage.

From his refined point of view, Rothenstein lamented the Hudsons' taste as they 'lived with the most forbidding furniture, the commonest pictures and china, the ugliest lace curtains and antimacassars'. Hudson, in a letter to William Canton, called his area a brick-and-mortar wilderness. By then the large semi-detached villa, built not much before in 1865, with a tower, had been converted into three flats. This tower made it the tallest house in the cobbled street. Hudson jokingly called it his 'Illimani', the Andean peak celebrated by Humboldt, as it was high and cold. In *Green Mansions* he runs through famous Andean peaks – Chimborazo, Antisana, Sorata, Illimani and Aconcagua – as 'names of mountains that affect us like the names of gods'. It was where he could seclude himself to write, as Montaigne did in his den and tower (and later Yeats).

The Hudsons lived unfashionably 'north of the park'. Compare that to Hyde Park Gate where Virginia Woolf was born or De Vere Gardens where Henry James lodged. But, at least, it was within walking distance of Kensington Gardens, and he walked there nearly every day. He would die alone at 40 St Luke's Road in 1922.

In a letter dated 4 April 1921 Hudson admitted to Violet Hunt that 'I was never in love with my wife' and that he had married because her singing voice moved him, but contradictorily, that Emily was 'the one being who knew me'. David Garnett, son of Edward, at the heart of literary London, saw Hudson from a boy's perspective around 1908: 'What impressed me most then was the gentleness and affection with which he [Hudson] addressed her [Emily]. There was a total absence of the aloof touch of bitterness he so often showed. It was clear that he loved her and wanted to make her happy.'

That aloof touch, though, does chime with other witnesses to Hudson's life who found they never really 'knew' him. Hudson's younger brother, his most intimate companion, had laughed at Hudson calling England 'home' before the *Ebro* had even left Buenos Aires, Bidding his brother farewell, Merriam added: 'Of all the people I have known you are the only one I don't know.' Emily may have been the sole person who knew him, but left no surviving record.

Hudson hated London, yet never chose to live in the country. Even before his arrival on the *Ebro*, he had been forewarned about London. Cranky tutor Mr Trigg loved reading and acting Dickens aloud to the six children at Las Acacias. Hudson knew the opening to *Bleak House* by heart (he had punned on a character in his letter-diary home in 1874) outlining the implacable November weather, mud in the streets, smoke from chimney-stacks, and especially smog. Dickens's London did not change much from the 'London' he created in the minds of countless foreigners until the Clean Air Act of 1956. Pea-soup fogs, caused by coal smoke and fog, would often prevent me going to school in the 1950s. It wasn't a joke. In one week in December 1952, 4,000 Londoners died. Hudson had travelled from an under-populated, non-industrialized country with enormous open spaces and a blue dome of sky and no pollution. All his later rambles were in areas that remained free of smog.

Hudson, like any immigrant, had to relearn everything from scratch. For example, he quickly grasped that scents on men were distasteful. A kindly, respectable barrister wanted 'to make an Englishman' of him and advised him to wear a silk hat, frock coat, tan gloves, neatly folded umbrella and above all, read *The Times* every day. That is, dress and act like a gent and you'll become one. But his friend howled when he covered his hanky in cologne water. No English gentleman would dream of reeking of scent! But Hudson had lived the life of a tough gaucho, hobnobbing with 'savages and dangerous whites', and wasn't scared of appearing effeminate. His hosts had confused the exaggerated scent of whores clinging to countless young men from English public schools and universities, so that scent became associated with immoral women. Hudson quickly learned to become respectable.

Chapter 8

Food and Carnal Cogitations

Maybe food gets to the heart of Hudson. He was brought up as a pioneer on the pampas, surrounded by semi-feral cattle, sheep and horses. Meat was the daily food of the gauchos. William MacCann, in 1848, was alarmed at their exclusive diet of beef and mutton, with mate, but no bread, no milk and no veggies. Grilled beef, the *asado*, is still Argentine identity-food. The tame Indians at Azul or the wild ones across the frontier only ate mare's meat. The delicacy was eating raw liver, still hot from the mares, with handfuls of added salt.

In *Idle Days in Patagonia*, we learn how the Indians would tie a horse by its hind legs from a branch, so blood would pour down. An artery would be opened in the neck and the blood caught in earthen jars. They would drink 'the abhorred liquid, hot from the heart of the still living brute'. Frontier life determined that you ate what was immediately at hand.

In fact, as he aged, Hudson verged towards vegetarianism. He felt revulsion at the thought of killing cows to eat them. 'Slay not' is the

Hipólito Bacle, *Carnicería*, 1839.

motto from *A Crystal Age*, repeated in *Green Mansions* as 'shed no blood, eat no flesh', the sentimental basis of English vegetarianism – killing animals as cruel. Cows, to Hudson, were 'gentle large-brained' creatures that 'caressed our hands and faces' with their rough blue tongues, more like a man's sister than any other non-human being. The 'very thought of beef' sickened him and eating it was killing it twice. This is strong stuff. My English 'step' grandmother was one of these sentimental vegetarians. She was a suffragette, had a blue butterfly tattooed on an ankle and was an intellectual (I've read Plato in her underlined editions). She taught me chess. She also converted her Norwegian husband, my 'grandfather', into not eating any more meat. He really took it to heart, made everyone do morning exercises, eat stale bread and even patented grass to spread over food. My family's not eating of meat came down from them.

Rubén Ravera, as director of the Hudson museum at Quilmes, had excavated under the birth hut and showed me photos of his discoveries, but not a bone from a cow or bullock. Hudson limited

Carlos Pellegrini, *El Matadero*, 1832.

cows to giving him his daily glass of milk. And drinking milk on the pampas was as weird as not eating meat, and still is. His family feeding habits made him a misfit. Hudson offers nothing about wild cows being more dangerous than bulls, as MacCann noted – they attack looking for you, with their eyes open.

An obvious source of his vegetarianism was his slow revulsion with the slaughter of cattle in Argentina. As a boy, barely a day passed when he didn't see something killed. A lasso was thrown over the horns, tendons were sliced by a gaucho with his *facón* and then throats were cut so that a great 'torrent of blood would pour from the poor tortured beast'. The gauchos would yell with glee and would ride the dying beast as its bellows subsided into deep 'sob-like' sounds. Around every shack on the pampas was a stench of rotting meat, drying hides and offal. Drying hides still hang over barbed wire around today's *puestos*. Dead cattle are still left to rot on the pampas, buzzing madly with flies and fought over by carrion birds. The smell makes your horse shy as it seeps into nostrils from miles away.

Hipólito Bacle, *Corrales de Abasto*, 1834.

The killing-grounds that supplied meat to Buenos Aires and that exported *charque* – jerked or dried beef – to feed Brazil's slaves particularly struck young Hudson. Slaughterhouses are well hidden today behind high walls and smells are eradicated. There's concern for the suffering animals and killing is quick, no longer a gory sport. With less adrenalin in the system, meat tastes better too. In taut, emotional prose Hudson evoked the Buenos Aires slaughter yard as a three- or four-square-mile field, enclosed with upright tree posts, with sheep and long-horned cattle constantly arriving in clouds of dust.

At Liniers, I witnessed a day's procession of lorries with trailers, each carrying forty squashed and shitting bullocks to the slaughter yards just outside Buenos Aires city limits. The vibrating lorries stretched out of sight as they queued in line. That dawn at Liniers over 17,000 head of cattle were slaughtered. In Hudson's time, hundreds of animals were killed every day in the gaucho's 'old barbarous way'. The gaucho's notorious indifference to death emerged from such continuous, callous killing.

Black-headed gulls, rats, pariah dogs and carrion hawk fed off what was left after hides were skinned and fat taken for tallow. So much spilt blood had 'formed a crust half a foot thick'. No English reader could imagine this scene, Hudson decided, as smells cannot be conveyed by prose. All around this *matadero* – the city had three such salting and killing fields – the local houses had fences and walls of piled animal skulls with horns protruding, covered in creepers. Darwin had noted this exact scene in 1833 and summarized it as horrible and revolting: 'The ground is made almost of bones; and the horses, and riders are drenched with gore'.

But Darwin never considered becoming a vegetarian as did Hudson, who witnessed this quotidian killing. One of the nastiest pieces written during Rosas's tyranny was 'El Matadero' (from '*matar*' to kill), an unfinished story by Esteban Echeverría, the earliest short story written in Argentina. It deals with the assault on an effeminate horseman in European clothes, riding on a European saddle and class enemy of the tough gangs who worked the slaughter yards. Just as he is about to be raped with a corncob he has a fit and dies. The story, set just before Hudson's birth in 1838, confirms that his memory was exact. Echeverría noted over 200 people, including many black women, stamping about on the muddy, blood-covered ground. A butcher worked on each carcass, without a shirt, his long hair matted and face smeared with blood.

Even in England these abattoirs stank, as Hudson's mentor Gilbert White complained. He planted four lime trees in 1756 to 'hide the sight of blood and filth' of his Selborne butcher's yard. Few who visit today's slaughterhouses could then sit down to eat meat. In fact, a good many would become instant vegetarians.

In *Idle Days in Patagonia* a cow escaped to an island, overrun by swine, at the mouth of the Río Negro. The single cow overcame her loneliness and adapted to the pigs and became famous. Then a gaucho from Carmen crossed over to the island and killed the cow. Hudson's comment in 1893 was that man was lower than the brutes. He ended his chapter, 'one does not sit down with a good appetite to roast beef or swine's flesh'. Hudson revealed his quirky diet, jokingly, to Henry Salt:

> I can eat sheep and pig and some other beasts, always excepting cow; also fowl, pheasant and various other birds, wild and tame, but I draw the line at wild geese. I would as soon eat a lark, or a quail or a nice plump young individual of my own species, as this wise and noble bird.

Hudson's family was not typical. In an essay 'The Potato at Home and in England', Hudson touched on his family's potato patch, with maize and pumpkins. He recalled his boyish delight at finding a wild potato as a tiny round tuber in the ground. He never travelled to Peru where they originated and can be bought in every shape and colour in any Andean market. Growing potatoes to eat divided the English from the Spanish settlers. Richard Lamb, in *The Purple Land*, was appalled at meat-eating waste. After breakfast, a wheelbarrow full of boiled and roast meat leftovers would be dumped in a heap for hawks, rats and gulls. They didn't even eat biscuits; as for a 'potato, one might as well have asked for a plum-pudding'.

Hudson recalled his family inviting a neighbouring girl to play with his sister. At dinner, there was lamb and potato. The girl ate the lamb and then, not knowing how to eat a potato, picked it up and dropped it in her cup of tea.

Yet, Hudson was proud South America had given the potato and maize to the world. When he arrived in England, aged thirty-two, he was shocked at the sight of his first English potato dish. A 'sodden

The Tero or *Vanellus Chilensis*.

mass of flavourless starch and water' was how he described mashed potatoes. He had always eaten them in their skin, as only peasants did in Europe. For Hudson, with his delicate digestion, the potato in its peel was his medicine: 'Nothing but potatoes for a day or two and I'm well again'. In his novel *Fan*, he ranted against 'that hideously monotonous mutton chop and potato which so many millions of unimaginative Anglo-Saxons are content to swallow' every day.

Far Away and Long Ago contains an elegy to his Argentine childhood food, made more vivid by his acceptance that as one became an old man smell and taste waned and digestion faltered. So his childhood cornucopia of four meals a day loomed larger than it should. The evening meal was a salad of cold sliced potatoes and onions, drenched in oil and vinegar. Breakfast was hot maize cakes eaten with syrup. They ate mutton, and gorged on eggs from chicken, goose, duck and wild fowl, with an occasional ostrich egg. Of the wild eggs, they mainly ate plover and lapwing, known as *tero*, an onomatopoeic call that still characterizes the pampas. Hudson hunted *teros* the native way, galloping past and noting where the birds rose (as they nest in grass). Once he collected 64 *tero* eggs. I found that seeing where the *tero* lays its eggs on the grass is actually far harder than he made it seem. I have walked up and down the enormous fields with these birds screeching and dive-bombing me and still not found their precarious nests. If there is one bird sound I associate with the pampas it's the *tero*'s irritating cry. Hudson also remembered cutting out honeycombs from the family beehives for breakfast – and being stung by bees. The peaches from their orchard ended up as peach pie or preserves.

Behind this litany of childhood food stood Hudson's hard-working mother, who fed six children and was a 'clever and thrifty housekeeper'. Thrift betrays the Puritan inheritance. And, of course, they all drank coffee, much to the scandal of the mate-drinking locals. Food is emotion from mother's kitchen. We are all conditioned by family tastes.

The meat-only diet of the traditional *criollos* did not coincide with Hudson's Puritan farm background, which ensured that he ate simply. Hunger and appetite defined healthy eating. In the futuristic society of *A Crystal Age* we have this sober, Zen-like philosophy: 'At night we sleep; in the morning we bathe; we eat when we are hungry.' Hudson may even have sided with George Bernard Shaw, who said, 'I am no gourmet, eating is not a pleasure to me, only a troublesome necessity.' By 1904, he told Algernon Gissing in a letter, that he scarcely ate any meat, and milk remained his 'favourite food'. He compared his late abstention from meat to Newman, who didn't touch it after reaching his fortieth year. He was referring to Cardinal Newman's younger brother Professor Francis Newman, who was an innovating vegetarian and president of the Vegetarian Society in 1873–84.

But Hudson was no teetotaller. He mentions drinking claret and ale while tramping England, even though his family were religiously opposed to alcohol. Mr Trigg, the family tutor, went off on weekends to down Brazilian rum, the only substitute for an Englishman's 'dear lost whisky in that far country', for 'at home there was only tea and coffee to drink'. In *Nature in Downland*, Hudson asserted that he was not an abstainer: 'Wine is among the kindly fruits of the earth which I appreciate, and failing that I can drink ale or stout', even though the 'perpetual swilling' of beer in Chichester was enough to 'turn the stomach of even the most tolerant man'. So sobriety, that difficult middle ground, was his philosophy, or as Ezra Pound put it about Hudson's literary style, his charm was his sobriety.

Hudson's diet was backed by the new 'cult' of anti-vivisection at the end of the nineteenth century. In 1885 *The Times* published letters lamenting vivisection. In 1891 Henry Salt, classic scholar, gave up being a top-hatted Eton master and founded the Humanitarian League to campaign against suffering in all sentient beings. He became a socialist and a vegetarian, growing his own food. He was an admirer and then friend to Hudson, as well as to George Bernard Shaw, Mahatma Gandhi and William Morris. Hudson wrote a preface to a Humanitarian League pamphlet on caged birds in 1911, and in 1918 penned another about a chained and caged linnet.

Salt met Hudson through the RSPB's Mrs Phillips around 1893. He'd reviewed Hudson's book on British birds, suggesting that Hudson jettison the dull scientific opening passage by Frank E. Beddard, FRS. Beddard worked as an anatomist at the Zoological Society and published on bird structure and classification. When Hudson met Salt, he chuckled with laughter over the putting down of Beddard. Salt called Hudson and Thoreau poet-naturalists. He'd invited Hudson to a conference he'd organized on Thoreau's centenary, but Hudson was too ill to attend and had Salt read his letter out aloud (I rejoiced to find this letter):

> Nevertheless I will stick to my belief that when his bi-centenary come round, and is celebrated by our descendants in some Caxton Hall of the future; when our little R. L. Stevensons are forgotten, with all those who anatomized Thoreau in order to trace his affinities and give him true classification – now as a Gilbert White, now as a lesser Ralph Waldo Emerson, now as a Richard Jefferies, now as a somebody else – he will be regarded as simply himself, as Thoreau, one without master or mate, who was ready to follow his own genius whithersoever it might lead him, even to insanity, and who was in the foremost ranks of the prophets.

Salt, who wrote studies on Thoreau, Melville, Shelley and Jefferies, thought Hudson a vague thinker and a man who could be petulant and perverse. He acutely summarized him as 'wayward and incalculable in his moods, never fully revealing himself, but always friendly and kind'. They would meet for a weekly tea at a vegetarian restaurant around the turn of the century, which Hudson rubbished as a 'vegetable place. I always had indigestion after dinner.'

After Hudson's death in 1922, Henry Salt and Bertram Lloyd, another keen birder and loather of hunting and war, honoured Hudson by deliberately following his footsteps in a walk near Shoreham. The overlap between bird protection, attempts to abolish cruelty to all animal life and anti-vivisection offered Hudson a quirky home in his later London years, summarized in Salt's title for his last book, *The Creed of Kinship*, 1935.

And so did vegetarianism. In 1870, the Vegetarian Society had 125 members, but by the early 1880s there were 2,070. The London

Vegetarian Society established itself in 1888. Gandhi had arrived in England the same year, read Salt's *A Plea for Vegetarianism* and then met him. He'd faced similar dilemmas to Hudson about how to become or at least behave as an Englishman. He took elocution lessons, learned to eat with a knife and fork, danced, read *The Times*, but refused to eat meat. As Ved Mehta noted, instead of mimicking an Englishman, he became more Indian. In fact, he read the Bhagavad Gita in London for the first time. Hudson often turned up at a London vegetarian restaurant, but didn't become more Argentine.

Hudson reflected on these London years in his tropical jungle romance *Green Mansions*, 1904. Old man Nuflo looked after Rima. He apologized for not giving Abel, the narrator, meat, especially when game was so tame and abundant in the woods. Nuflo had apparently become vegetarian thanks to Rima. He said: 'For us, señor, every day is fast-day – only without the fish. We have maize, pumpkin, cassava, potatoes, and these suffice.' For Rima, a vegan, preferred wild berries. So because he loved Rima, Nuflo shed no blood and ate no flesh. Abel was incredulous until he discovered that Nuflo had remained a secret meat-eater.

Hudson at home on the pampas was familiar with garlic. But in England it was seen as 'such abomination to the English palate', the worst of foreign excesses. However, Hudson would try out any food, in the manner of a trained anthropologist. A good example is chewing gum. During summer in Patagonia, recovering from his wounded knee, Hudson tried *maken* from the gum of a juniper-bush. The natives roll it into pellets and melt it over water to form thick, putty-like drops. He had to practise to chew it, and often made a sticky mess in his mouth, but its resinous, refreshing taste would last for a week. Hudson found that chewing this gum cut the need to smoke and preserved teeth with a whiteness unseen anywhere else. *Maken* chewing was common all over Patagonia. He enjoyed teasing his English reader by boasting of having tasted vizcacha, the pampas rodent, while writing an essay in 1872 before he left Argentina:

> It is a very unusual thing to eat the vizcacha, most people, and especially the gauchos, having a silly, unaccountable prejudice against their flesh. I have found it very good . . . the young animals are rather insipid, the old males tough, but the mature females are excellent – the flesh being tender, exceedingly white, fragrant to the nostrils, and with a very delicate game-flavour.

Alas, he was no cook. In his later more affluent years, he would lunch out on Tuesdays at the Mont Blanc at 11 Gerrard Street. There's a plaque on the site stating that many leading writers met there, including Conrad and Galsworthy, although Garnett, who organized these literary luncheons, claimed the former rarely turned up. Hudson isn't named on the plaque.

He also lunched under the dome at Whiteleys, the department store in Bayswater, where he invited Helen Thomas just before his death. Violet Hunt also recalled lunching there when the aged Hudson chose spinach, cheese and coffee. She claimed he lived on air, and once just 'eked out' two potatoes and didn't touch the cutlet. He was frugal.

In a late piece called 'My Friend the Pig' Hudson mocked vegetarians who desisted from meat while ill, but when healthy stuffed themselves with roast pig. He befriended a pig in a stinking muddy pen, learned what the pig thought, fed it an apple then bunches of elder berries, until it was carted off to be slaughtered. He savoured the idea that someone would eat a rasher tasting of elderberry. English vegetarians are not like Thoreau, satisfied with his handful of rice; they're not frugal, and don't treat fellow animals with love.

In Hudson's anonymously published *A Crystal Age*, 1887, the protagonist Smith, typical prejudiced Londoner, stumbles while botanizing, into a valley of the future where nobody has even heard of London or England or Queen Victoria. Hungry, he is invited to the evening meal to find a plate of whitey-green, crisp-looking stuff that is cold and bitter (a bit like endive), followed by vegetable dishes, fruit juice, crushed nuts and honey, but no meat and no booze; 'the delicious alcoholic sting was not in them'. Then Hudson jokes that what Smith really hungers for is a 'haggis', but that is 'that greatest abomination ever invented by flesh-eating barbarians'. So Hudson defied the Hogarthian 'O the Roast Beef of Old England' excesses, while not allowing himself to be pigeon-holed as a vegetarian.

From attitudes to food to the taboo topic of sex – how did Hudson navigate his way in the sexual field, those 'carnal cogitations' that so alarmed John Bunyan? He confessed his attraction for little girls: 'seven is my limit. They are perfect then'. Today his girl-love would be suspect. But, as with Alice Liddle and maths lecturer Charles Dodgson, or John Ruskin's Rose La Touche and back to Dante's Beatrice (rather than Nabokov's Lolita), girls before puberty embodied an ideal of beauty and purity. Hudson's close family group of five siblings, with two of them sisters, fostered this sexless adoration. Hudson and Emily were childless, and at one time schemed of adopting a daughter. As he wrote in a letter to poet William Canton, 'we are great lovers of children but have none of our own'.

In August 1903 in an unnamed town on the Norfolk coast, possibly Wells-next-the-Sea or Cromer, Hudson noticed two little girls on the beach:

> They were dressed in black frocks and scarlet blouses, which set off their beautiful small dark faces; their eyes sparkled like black diamonds, and their loose hair was a wonder to see, a

black mist or cloud about their heads and necks composed of threads fine as gossamer, blacker than jet and shining like spun glass – hair that looked as if no comb or brush could ever tame its beautiful wildness.

He adored their instinctiveness, their wildness, and a mass of hair was an added attraction. These little girls could have been birds (slang highlights this link). Hudson would follow these two girls, try to talk to them, but felt like a dodo chasing hummingbirds. He loved the purity of their voices. By their black diamond eyes as well as their vivacity, he realized they must be Iberians.

Hudson told a childhood story about a neighbouring girl, aged about eight, from the dreaded Barboza family. He passed her every day as she rode bareback like a boy, working with sheep and cattle. Hudson, old and in bed writing, admitted: 'I can still see her now at full gallop on the plain, bare-footed and bare-legged, in her thin old cotton frock, her raven-black hair flying loose behind.' She was extremely white, like alabaster, and lean and serious. To young boy Hudson, on the verge of puberty, she was beautiful, with a cloud of hair on her. But he was too inarticulate to ever approach her. Then Hudson left and never saw her or her queer tribe again, but the image of her in his mind never lost a 'certain disturbing effect'. That final confession: Hudson was excited by her bare-backed riding, her hair and white legs and face, as any boy approaching puberty would be.

Fifteen-year-old Marta from Hudson's Indian tale 'Marta Riquelme' fits into his type. She was Spanish, with white skin slightly darkened by living in the tropics and eyes with a violet tint, and her hair, 'the crown of her beauty and chief glory', was long and shining gold colour. Hudson fixates on hair in his girls. In the tale the priest narrator falls for her and struggles with his lust.

All men struggle with lust, but Hudson was deeply reticent in a reticent age. Yet, his first thirty-two years were spent in a primitive society where physicality was upmost. Tough gaucho work, on horse, branding, herding and sleeping rough determined Hudson's traits. He would have seen stallions climbing the backs of mares, rams on sheep and dogs on bitches in constant, instinctive copulation. Gauchos might settle with their *chinas*, have children and roam the plains freely, but outside every village there are still brothels and a cult of masculinity that exalts phallic power. This attitude was intensified during his military service on the frontier, outside society's norms, where in William MacCann's words, 'the tone of sexual morality is low'. Nothing is recorded about Hudson's sexual activities. He was married from 1876 to Emily. Can I assume that this was a sexual relationship, that they lived in separate rooms and that Hudson's Puritan upbringing had instilled a strict morality about extra-marital affairs?

Hudson reverts to conventional love plots in his fiction, with bouts of passion, inevitably doomed. Perhaps Hudson was monastic or even asexual? Did he sublimate his sexual drive into field observations? What about his sexual fantasies and secretive masturbation? Onto this biographic blank, a writer on Hudson's sexual life simply projects his or her own sexual fantasies. Publicly in letters, Hudson stressed how he and Emily became friends; she was a companion, 'more than wife', and that she was the sole being who really knew him. Her death left him 'very much alone'. But sexual attraction has been carefully eliminated.

Of course, there's his mother-love and love for little girls and a hair fetish, but common sense forces me to assume that Hudson's sexual drive was 'normal'. From his writings, Samuel Looker assumed Hudson's 'maidenish reserve', while Felipe Arocena, a recent academic biographer, assures his reader that 'we know that he had a number of lovers, and at least one of his affairs lasted for many years'. Neither version is certain. Only a novel could explore his sexuality, by reinventing it.

In *A Crystal Age*, Hudson's protagonist is sexually re-educated. A typically outdoor Englishman, Smith relearns how to love. He falls for Yoletta but is puzzled by her lack of physical response. There's a telling scene while Smith and Yoletta are out on a walk. He grabs her hand, kisses her palm, then each finger-tip until she asks why he's doing this. He continues kissing her cheek, then her chin:

> 'Yes; but you are keeping me too long. Kiss me as many times as you like, and then let us admire the prospect.'
>
> I drew her closer, and kissed her mouth, not once, nor twice, but clinging to it with all the ardour of passion, as if my lips had become glued to her.
>
> Suddenly she disengaged herself from me: 'Why do you kiss my mouth in that violent way?' she exclaimed, her eyes sparkling, her cheeks flushed. 'You seem like some hungry animal that wanted to devour me.'

Smith's animal passions are slowly re-educated, sublimated, into a mystic passion and mother-worship.

Whatever one thinks of this odd novel, read as a confession it suggests that Hudson, in 1887, aged forty-six, questioned the centrality of sexual love in human nature. In his preface to the second edition, Hudson admitted that when he wrote he did impart something of his soul to the paper. Why then envision a future where the sexual urge is redeemed? Paul Theroux argued that Hudson ranted against the sexual impulse, that contemporary man is oversexed. The most perceptive critic of this novel, Desmond MacCarthy, in the *Listener* in 1932, found Hudson something of a mystic, but didn't want to make too much of that word. In his utopia Hudson had chosen love for nature over love between the sexes. 'This beautiful and moving book is also,' he wrote, 'a picture of what this loss entails.'

In a letter to Edward Garnett of 10 June 1913 – Hudson was seventy-two years old – we have the following opinion on D. H. Lawrence:

> Lawrence is all right no doubt, only I've had so much to do with flesh that his insistence on it – its warmth or hotness, its colour, the curves of it, and the kisses pressed on it with hot, wet lips – goes against me. By and by when he grows out of this stage, which often comes to a young man who has repressed all his sexual instincts from religious motives, until he gets himself suddenly free of them, he may do some really good work.

What does he mean that he has had so much to do with flesh? Is that an old man's boast? By flesh, did he mean his field naturalist/gaucho experiences of natural animal sexuality? Or does flesh refer solely to women? Just as intriguing is that link between repression of sexual instincts and religious education that was Hudson's as much as Lawrence's. One can only guess at the untold stories there. Hudson deemed *Sons and Lovers* a very good book, except when Lawrence relapsed into 'neck-sucking and wallowing-in-sweating flesh', an obsession and madness that he will outlive as so many other writers have. So it's clear: one thing is writing about it, another is having lived it. Hudson never wrote that way, but did he live that 'madness'?

In 1916, Hudson wrote to Garnett about *A Crystal Age* and the 'sexual passion', its central topic. There can be no peace, no rest, until the sexual fury has burnt itself out. He admitted that this future was remote, as the violence of the sexual rage does not abate and burns as fiercely as it did 10,000 years ago. There is no end to prostitution, 'millions and millions of women in that stage just to satisfy men's ferocious desire – not only of the young unmarried men, but of all men as an everlasting protest against the law that forbids a man to have more than one wife'. So Hudson held strong feminist views about man's animal urges and how it sours the spirit.

I sense a lifetime struggle with his own sexual instincts. I glimpse a man bent on recuperating instincts as his rebellion against artificial urban society, but dismissing the sexual urge.

However, there's an apparent contradiction and it concerns Linda Gardiner. She was secretary to the Society for the Protection of Birds from 1900 to 1935, when it shared offices with the Zoological Society at 3 Hanover Square and then at 82 Victoria Street. She also ran the society's journal *Bird Notes & News* from its first issue in 1903, and edited Hudson's posthumous booklet *Rare, Vanishing and British Birds*, 1923. In all, she worked for the RSPB for thirty-five years. As a seventeen-year-old, she wrote a column on birds for her father's newspaper. Before getting her job, she had published five thrillers between 1895 and 1899, all republished by the British Library. So she also knew how to write from the inside.

She shared Hudson's disgust at the plumage trade, calling it a 'barbarous fashion' (*Evening Standard*, July 1891), and wrote letters to the papers on feeding birds on windowsills (like Hudson himself). She defended traditional values for woman, but had the original idea to install migrating birds' rests on lighthouses. She sat on the committee that chose Epstein's statue of *Rima* as the Hudson Memorial in Hyde Park.

A cache of thirty-eight extraordinary letters from Hudson to Linda that run from 1901 to 1921, kept in Manchester Central Public Library and long hidden from view, has been transcribed and published by Dennis Shrubsall from photocopies held by the RSPB.

Linda Gardiner in later life.

Just the kind of letters Hudson would dearly have liked to burn, as not one of her letters to him has survived. Why did Linda Gardiner keep them? In a late letter to her, Hudson explained how he had destroyed over 2,000 letters, but asked her to save those from his RSPB friends Mrs Phillips and Mrs Hubbard, and sell them in the United States, but not in England, as he didn't want his literary friends to scribble about him. So did Linda Gardiner keep these letters for financial reasons? Clearly not, as her literary heirs refused an earlier biographer, Alicia Jurado, permission to read them. I believe Linda Gardiner preserved these letters from the bonfire because she wanted to be remembered as his true love.

Morley Roberts hinted that Hudson may have been 'incapable of the passion which may wreck a man', and certainly Linda Gardiner doesn't feature in Roberts's early biography. Years later, in 1937, Roberts wrote to Dent, the publisher: 'I didn't deny that H. was a strongly sexual man. But he never spoke of his affairs, how many they might have been.' He added, 'there was never a cleaner-minded man than him'. He reiterated that Hudson wished to 'remain mysterious, not to be understood'. Reading these surviving letters to Linda, I bear Roberts's words in mind about *not* wanting to be understood. The letters do not tell a clear story, and they're one-sided.

Hudson scrawled that he craved for Linda, but 'dare not even touch your lips'. He rebuked her often for not opening her heart to him, for not talking about her feelings, for being too reticent. He obviously wanted her to correspond, which she didn't seem to want or be able to do. In a letter of 1902, Hudson confided, in a hectoring, seducer's tone about love: 'For love, dear, is a bodily appetite as well as a spiritual, and it has no cure.' He cited 'the hidden passion, the swifter pulse, the brightening of the eyes . . . the flower's fragrance that comes to a woman's sweet lips'.

By June–July 1902, some barrier has been crossed for Hudson refers to missing 'your dear warm fragrant lips'. From the 'Miss Gardiner' of 1901 we now have 'Dearest Linda' in June 1902 (and by 1905 his favourite name for her 'dearest Ethelind'). Yet 'lips' is as far as he seems to have got, for he wrote 'even if I had – in the old English sense of the word or phrase—*known* you?' But then by 29 August 1902 she suddenly seems to have written 'fully, freely from your heart'. He offered to send her a photo, and expressed love and tenderness. On 17 September he told her that 'we are doing wrong to no one so far as I know' (even if he was married to Emily). But then he assured her that he used a secret mailbox at a stationer's in Bayswater near his home. He continued to write to his wife Emily, but never once mentioned Linda.

So sometime in 1902 'something happened' beyond the usual meetings for tea. Hudson reminded Linda 'and you are dear to me, your body as well as your mind'. Then a later 1902 letter concerned a Shadow that came to him at night, respecting her virginity. This

Shadow 'gives me her body to kiss and her sweet breasts for consolation'. The Shadow is not Jungian, but that other, daring, nocturnal
self that dismissed moral quibbles and social codes.

Then some time in early 1903 Hudson's tone has changed. He
doesn't want Linda to be unhappy, patiently waits for a letter and
'for a good few hours to be together'. Then there's another twist for
on 28 December 1904 Hudson consoles Linda for having broken
some barrier:

> but it is wrong of you to lie awake and think you must have
> been mad. Think rather (with astonishment if you like) that
> you must have been sane. For surely it is ordained that we
> must love, and you then have followed nature and the right
> when you overcame second nature and self and gave yourself
> to one who loved you.

How Hudson would have loathed being caught *in flagrante delicto*.
These letters raise questions of biographical prurience. But he's
clear. Linda Gardiner gave herself. But what did she give? Here we
descend to anatomical details. Did she 'surrender' her virginity, or
just climb into bed? Hudson immediately told her that he was going
to 'Brighton this weekend'. Wasn't Brighton synonymous with adul-
tery? Or was it an innocent reference to writing *Nature in Downland*?

Whatever happened, by January 1905, the 'affair' had cooled down
and Hudson found it so strange that they failed to understand each
other. Some time in 1905, he wondered if in fifty years time anybody
would know 'this secret of ours' and wrote in the past tense as if it
was all over: 'my friendship with you, and more than friendship, was
the sweetest and best and most purifying influence I had known', as
if 'friendship' didn't have a sexual element. Hudson was well aware
of the 'disparity' in their ages and health, but he wrote on 13 May

1907 that he didn't want the relationship to slip back into just friend-ship. By then what could have been physical passion had reverted into 'old dear memories and vain longings'. The 'dream demon has ceased from its persecutions', he wrote a little earlier in 1907.

So, whatever the real story was, from the surviving letters you are left in the dark. Letters are a continuation of a dialogue, but so much remains outside the handwritten words. But no doubt that Hudson had been smitten with love. His last letter to Linda was dated 16 August 1922, a few days before he died.

These letters turned my view of Hudson upside down. He seemed to exploit Linda as a sixty-year-old macho and she resisted in a spinsterish way. Her moral values were strong, the fact of her public position in the RSPB, the possible scandal, his wife Emily. Morley Roberts thought that Hudson did 'love' his wife. After her death, Hudson told Violet Hunt in a letter that he didn't. It seemed like a marriage of convenience, to avoid loneliness and exile, but that's not accurate either. She was also an avid novel reader who read his novels in draft and maybe even corrected his English. In these letters to Linda Gardiner, Hudson sounded like his character Smith from *A Crystal Age* or Richard Lamb from *The Purple Land*. The language is cliché-Gothic, over-heated, too sentimental. Today, it might be too Mills and Boon.

It boils down to a basic question: did William and Linda go to bed together or not? My guess is that they didn't; it's also the conclusion of Dennis Shrubsall. Ruth Tomalin, a discreet biographer, called it a 'tormenting affair', without going further into details. In Linda Gardiner's obituary on Hudson in *Bird Notes & News* it was a 'close friendship'. Hudson deemed it the best thing that happened in his life. In fifty years' time, Hudson guessed in a letter, readers will decide that their relationship 'was the sweetest and best and most purifying influence I had known'. I could find only one photo of an aged Linda, accompanying her obituary in *Bird Notes and News*, while digging about the RSPB archives.

After Hudson's death, Linda Gardiner's long obituary wasn't personal. She recalled him as a reticent man, who in old age showed little signs of age, with an alert mind and faculties untouched. In 1902, Hudson was sixty-two and quite capable of making love, and certainly of falling in love. Linda was twenty years younger. So Roberts was wrong: Hudson suffered the passion that wrecks lives. But it probably remained verbal, in secret letters or over furtive cups of tea. Linda Gardiner wasn't at Hudson's funeral in Worthing, but sent flowers. He left her £100 in his will 'in gratitude for her friendship', which probably didn't have a double meaning.

I found in Buenos Aires a letter from Linda Gardiner to Edmund Gosse, from the Royal Society for the Protection of Birds, typed on 25 June 1923. She reiterated that Hudson did not want his letters to become 'copy' for any of his literary friends who might want to write about him. She was sure that Gosse would not use them for such a purpose. But after reading her letters I became aware of an alternative version, that it could be a mistake to think that a couple like Hudson and Linda Gardiner, from another age, would not have obeyed their animal instincts. It would be like thinking that your grandparents never made love. Nevertheless, I dismissed this supposition. Linda Gardiner died unmarried in 1940.

In 1923 Linda Gardiner updated and posthumously published Hudson's notes on disappearing or extinct nesting birds in Britain. He had been weary and old, preoccupied with his last book *A Hind in Richmond Park*. She restated his passion for 'the mystery and beauty of life' and how he hated gamekeepers, greedy collectors and louts with guns. She had access to his notebooks in pen and pencil that were 'extremely difficult to decipher'. And added, with irony for us today, 'even for those intimately acquainted with his handwriting'. I imagine her sitting at 82 Victoria Street struggling lovingly with his illegible handwriting. She almost boasted 'that I probably knew more of his mind in regard to this particular subject than anyone else knows'. But most revealing of her love is the date, 22 August 1923, which she placed after finishing her foreword, for it was exactly a year after his burial in Worthing. One last point is that nobody ever guessed there was anything improper in the many RSPB committees they sat on, such was their skill in dissimulating.

In Spanish, Linda and Rima rhyme and translate as 'pretty rhyme'. To read biography back into fiction, I know, is a desperate attempt to mine suppressed facts. Hudson wrote fiction for money and tended to despise it. But what's crucial for the Linda/Rima link is 1904, the date of publication of *Green Mansions*. Right in the middle of Hudson's apparently unrequited passion for Linda Gardiner.

In between arranging secret trysts or accompanying Linda at RSPB meetings, Hudson wrote his tropical jungle romance. Its essential theme is that protagonist Abel cannot possess Rima. She's too ethereal, too mythical, too idealized. He can only mourn her death, her scorched bones still warm in his hands. The novel is about an impossible love for a fragile hummingbird kind of woman, Rima, too frail for sexual possession. The nearest Abel got was to kiss her. 'The shame was as nothing in strength compared to the impulse I felt to clasp her beautiful body in my arms and cover her face with kisses.' Abel felt 'sick with desire'. Later, he kisses her again. When Rima dies, burnt alive by her enemies like a witch, she becomes a 'visionary' Rima, just as Linda had been. Just like the writer of the letters to Linda, Abel picks up the bones, the sacred relics, he kisses each fragment, collects the bones and takes his 'treasure home'. This mental act of memory was all that was left to a frustrated Hudson, who got no further than a kiss.

In 1909 a critic, E. R., guessed that *Green Mansions* revealed the 'secret of Mr Hudson's personality' as a man suffering a 'passion that cannot be granted'. E. R. met him once at Golders Green, when Hudson had come to see if the final chapter of *Far Away and Long Ago* should be excised. It was 1916 and Hudson was seventy-five, with an 'alert, erect figure', an 'aquiline nose, keen-eyed, bird-like, well-weathered face'. He hid behind his impassive mask.

I'm no decoder of secrets, but will make three guesses. The first is that, in Hudson's inner world, only he thrilled at the private associations of that rhyme Linda/Rima. Second, Linda could not be possessed and Hudson was ageing, losing his virility, like his character Abel, 'beyond all that fiddle'. Third and most crucial, was that Linda and Rima were strict vegetarians who couldn't tolerate sexual violence, machismo or possession because both wanted to be free. Killing animals was a kind of murder, we read in *Green Mansions*. Linda Gardiner wrote in a RSPB pamphlet titled 'Birds and Boys' that 'true naturalists' didn't kill. Rima taught that 'Shed no blood and eat no flesh'. So my QED is that Rima is Linda. While writing his novel between 1902–4, she haunted him.

However, I juggle with Morley Roberts's version that Rima could also be his wife Emily, very short and elfin-like in her youth. Violet Hunt also thought Rima might be Emily as, despite being so short and needing a stool for everything, she sang divinely. If Emily was Rima, Epstein did her justice. But neither Roberts nor Hunt knew about the Hudson/Gardiner liaison. It was completely secret. Alas, biography confirms the uncertainty principle.

Chapter 9

Birds of a Feather

Although he saw himself as a field naturalist commenting on all natural life, it was as an ornithologist that Hudson gambled his reputation. But he had to divorce himself from the scientific ornithologists, as how he wrote about birds counted more than mere listing of attributes. He'd done his taxonomic slog-work on Argentine birds. I've seen his handwritten lists, but always sought a more emotional understanding.

He began like any boy from a rural area and stole eggs in nests and shot birds, but soon he'd discovered a vocation. For him, birds exceeded in beauty all other animals, so there's an aesthetic element to his emotional identification. But even more attuned to his later animistic philosophy was the notion that birds embody an intensity of 'life so vivid, so brilliant, as to make that of other beings, such as reptiles and mammals, seem a rather poor thing'. He thrilled to a particular bird intelligence that brought him into a closer harmony with the environment. He observed the 'countless little acts which

Goldfinch by A. Thorburn.

result from judgement and experience and form no part of the inherited complex instincts'.

My favourite example concerns a moorhen he observed in a west London pond. The scene couldn't be more mundane. The moorhen was feeding on the banks when it spied another bird. It lowered its head and rushed it. Then did this again, but the other bird didn't budge. Finally 'it began to walk *backwards*, with slow, measured steps . . . displaying, as it advanced, or retrograded, its open white tail' and peering back to see how this tactic worked. Hudson couldn't guess whether its motive was anger, love or just fun. At the heart of his writing is a modernist credo: 'The bird-watcher's life is an endless succession of surprises.'

Beyond this aesthetic and emotional source to birdwatching was a drive to understand birds. By patiently observing them, he could identify with them. He was a bird psychologist, constantly project- ing himself into birds' minds, whether a vulture spotting carrion or a migrating swallow. He mastered their language of sound and motion, which 'tells you what they feel and what they mean'.

Then came the dream of discovering a new species. Before he had left for England in 1874, Sclater had already attributed to him two new Patagonian bird species. This was a brilliant start to a promising career in London. But in England, with its long tradition of bird observation, there wasn't a hope of discovering another new species. It made his Argentine days, in retrospect, more exciting. He couldn't compete with English ornithologists and didn't return to Argentinian observations, except in memory.

Most striking is how he retained bird images in his memory. Once again, restless city-dwellers have not really *seen* birds. They are too preoccupied with everyday life. He wrote: 'If they had ever looked at wild birds properly – that is to say, emotionally – the images of such sights would have remained in their minds.' By recording bird-images emotionally in his mind, he can hold onto 'my treasures, in my invisible and intangible album'. His birds become a 'permanent possession'. You begin by patiently watching, mind emptied, being moved emotionally, then record what you've observed in memory via images and words. He thus defies time by hoarding images of birds.

At the end of his life, he returned to his notes and memories of birds seen in his childhood Argentina. 'South America can well be called the great bird continent', he wrote, for more species existed there than in England. He especially remembered the golden plover. He would mount his pony and gaze with excitement on a *floor* of birds, like a sea, vibrating with song. Hudson found it 'indescribable' and 'unimaginable'. Above all, the mystery of migrating geese gripped him. Lying in bed in his hut on the pampas, he would listen to their sound coming from the sky so that it lived on inside him, as vivid as but an hour ago. Bird experiences were filtered through emotion and memory and emerged in writing.

Finally, Hudson was convinced birds were disappearing. Something was coming to an end. The book he wrote with Linda Gardiner captures the melancholy mood behind 'rare' and 'vanishing' birds. Hudson's lament that 'the beautiful has vanished and returns not' summarizes a lifelong pessimism.

Hudson was a doer, not a talker. His family Puritan background had instilled this practical bent. Nobody on his small farm sat about idling. His family faith, epitomized by his reading of John Bunyan, did not end 'in word or tongue' but 'in DEED and TRUTH'. So Hudson dedicated his insanely active English years to the protection of birds. This history of nature conservation has been well told by Ruth Tomalin. I appreciate this selfless work, but find it a bit dour. Nevertheless, he was in at the start of the Society for the Protection of Birds.

Hudson met Mrs Edward Phillips in 1889 and was invited to her lunch followed by discussion, known as the Fur, Fin and Feather Folk afternoons in Croydon. They remained friends for life and a one-sided copious correspondence exists. He was one of the few men in these early meetings, but doesn't seem to have worried the women. He was without sexual innuendoes and preserved an asexual, courteous front. At these Croydon lunches he met Mrs Frank Lemon who had set up a Society for the Protection of Birds in 1889 in Didsbury, Manchester, and she too became a life-long correspondent.

After the Croydon and Manchester groups joined up in 1891, Hudson actively supported the fledgling society protesting against the trade in the soft under-pelt of the great crested grebe. At the start in 1891, he wrote a pamphlet with a distribution of 5,000 copies. He then wrote its third pamphlet the same year on ospreys, egrets and aigrettes. And three more pamphlets followed, all with a vivid, coloured illustration of some bird. In pamphlet no. 73, 'On liberating Caged Birds', Hudson typically boasts and captures his

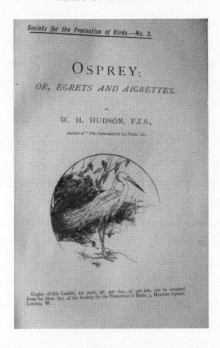

Society for the Protection of Birds.—No. 3.

OSPREY;

OR, EGRETS AND AIGRETTES.

BY

W. H. HUDSON, F.Z.S.,

Author of " The Naturalist in La Plata," &c.

Copies of this Leaflet, 1d. each, 9d. per doz., 5/- per 100, can be obtained from the Hon. Sec. of the Society for the Protection of Birds, 3 Hanover Square, London, W.

activist's doing: 'I know a great deal about birds, having been observing them all my life, and have also a good deal of experience in liberating them'.

The society fully channelled his rage against the use of bird feathers in ladies' hats, caging birds and killing birds in cruel ways with lime. He dedicated his energy to helping, with letters to *The Times* (one on the trade in bird feathers was reprinted as a RSPB pamphlet no. 12) and another to the *Observer* on seagulls in London in 1921. In 1894 he published an important ecological pamphlet on vanishing birds, edited into a short book just after his death. Also in 1894, he was appointed Chair of Council, but was replaced in 1895, probably because he 'had no aptitude for administrative work' in Mrs Lemon's words. One of his last official acts in November 1895 as chair was signing and writing 10,000 letters to church ministers about the society.

One of the 'best things Mr Hudson ever did for the Society', Mrs Lemon thought, was to suggest a replacement Chairman, who then stood for forty-seven years. Hudson remained on the Committee and Council until 1921, shortly before his death. He also joined the Watchers' committee and was secretary to his local North Kensington branch. On 14 April 1900, a letter to *The Times* titled 'Kew Gardens and Old Deer Park' was reprinted on a single sheet and folded into four. It was distributed free by the society, protesting successfully against the building of the National Physical Lab in the Old Deer Park.

I find it moving to imagine Hudson amongst these women, plotting campaigns against feathers in hats, his 'obsession' according to biographer Alicia Jurado. Mrs Lemon, his friend, found him always impeccably dressed and well-mannered, though serious and distant. She glimpsed a 'kaleidoscope' mind behind this mask. She noticed that his 'capacity for immobility when watching birds' was 'phenomenal'. His only movement was the flicker of an eyelid. He may have been aloof, with a chameleon-mind, but was 'very loveable'. He'd discovered that the unique English love for birds had given him roots.

As well as friendship with Mrs Phillips and Mrs Lemon, the society offered him a meeting place and agenda with other like-minded friends like the Ranee of Sarawak and Mrs Hubbard, an illustrator to some of his works, whose many letters he instructed in his will to be spared and sold abroad to raise funds. The bulk of the society's work was done by women, and they appreciated Hudson and surrounded him with affection as he'd attended most meetings.

One of the few males in this association was Viscount Grey of Fallodon, a busy Foreign Minister for eleven years in 1905–16 who led his country into the Great War. Through his bird-work, Hudson had reached the elite of England. Grey's tribute to Hudson on his death,

printed as a handout kept among Jorge Casares's papers, claimed he
had a 'permanent place in English literature'. He suggested Hudson
provided 'the pleasure of contemplation', especially for someone
like himself 'on the rack' of public affairs. He met Hudson 'fragmen-
tarily', as he was always in a hurry. He wrote that for many like him
going back to nature was a solace, that 'it is good to be alone with
nature sometimes, to men like W. H. Hudson, it is essential'. In a
letter to his first wife Dorothy in 1893, Grey writes that reading
Hudson after busy, public work in London made him sad as he was
such a long way off from that peace with nature, adding: 'I shall read
more Hudson tonight and store it up for you like honey.' But there
had been more than an ornithological contact with Lord Grey.

He lent Hudson his secluded fishing bungalow in Hampshire,
built in 1890. It was near a station, down an avenue of lime trees,
with corrugated tin roof and trellis with roses and honeysuckle, by
water meadows on the Itchen river. In 1923 it was burnt down and
now only a chimney stands. It was a *locus amoenus*, a patch where
time stopped. William Cobbett, in his *Rural Rides*, 1853, also picked
on the Itchen valley as 'certainly one of the prettiest spots in the
world', with meadows, pond and wild ducks. Hudson had merged
with this tradition, chameleon-like, of countrymen like Selborne's
White and Cobbett, and pretended to be another ruralist. It was
some feat for an immigrant.

Hudson wrote an obituary in the *Speaker* for Lord Grey's wife
Lady Dorothy, aged forty-one, killed when she fell out of a dogcart
at their baronial home in 1906. He recalled his first meeting with
her, her original mind, not uttering customary phrases in her chat,
being transparently honest, a 'native, so to speak, of my world'.
She would dismiss her servant to be alone on the Itchen river.
Hudson thought 'there are few persons who can endure solitude'
and who desire a 'simple life'. According to Michael Waterhouse,

Dorothy and Lord Grey lived an intimate *mariage blanc* as she had an aversion to children and refused physical contact. There were rumours that Grey hid a long affair with Pamela, who would become his second wife. Dorothy adored and recited Wordsworth and contributed to the nature diary both kept (*The Cottage Book*), where Hudson is mentioned. He connected with Dorothy's sexless sensibility and her independence from the world, even though married to a hectic Foreign Secretary. Hudson had caused him no jealousy.

Hudson claimed the Greys' fishing cottage, close to Itchen Abbas village, was one of the most 'refreshing places in Hampshire', almost in England, in his *Hampshire Days*, 1903. He added, 'I love the Itchen', which he discovered in 1900 (Winchester was the first cathedral he saw on arrival in England). There was no path to the cottage, which was hidden by lime trees and creepers, a 'delectable', even 'ideal' spot. He escaped London's dusty desolation and intolerable heat that July 1900 for the whole summer, just as the Greys escaped their politically hectic life there. Typically in 1903, after biking there, Hudson didn't mention in print that the Greys were his hosts.

Lord Grey co-founded the RSPB in 1893 and was its vice-president in 1909. His wife Dorothy became a life member in 1895. She wrote to Hudson: 'I have no feathers except ostrich feathers for the last ten years and have induced several people to give up aigrettes.' Grey wrote education bird pamphlet no. 5 on kingfishers and later the sentimental *The Charm of Birds*, 1927. He would give his name to the Edward Grey Institute of Field Ornithology at Oxford, where David Lack was director, 'the most renowned ornithologist ever'.

In 1903 we get a glimpse of the kind of footwork Hudson carried out for the society. He had rambled through eighty villages including Marlborough and talked to several county councils, all to get a new bird protection order. He wrote letters to General Purposes Committees and asked the local MP for Bristol, Henry Hobhouse, to put this order before his House of Commons committee. Hobhouse was at first unpleasant, but changed his mind. It was tedious, self-sacrificial work. The following year, 1904, the society was granted its royal charter (RSPB), and Hudson and the society kept up the fight for the Plumage Act, which did not pass through Parliament until 1921. In his will, Hudson donated his worldly goods and royalties to the society and especially instructed it to print leaflets on birds for children in village schools in an 'anecdotal' style, which Hudson himself had worked on to make himself so readable.

I found a seven-paged, handwritten lecture donated by Philip Gosse, the poet Edmund Gosse's son, to Jorge Casares, dated 29 October 1919. There's no name to it, but it's obviously written by Gosse himself (later I found it printed in his *Traveller's Rest*, 1937), and given to Casares when Gosse travelled out to Buenos Aires to give a casual talk on Hudson. It catches Hudson attending his last meeting of the RSPB. Gosse opens: 'It was at a council-meeting of the Royal Society for the protection of Birds, held in London October last, that I saw for the first time, and what proved to be the only time, the naturalist and writer W. H. Hudson.'

Hudson's poor health kept him wintering in Cornwall and thus seldom able to attend. He turned up late on this occasion. A door opened and 'there entered a figure which could not fail to arrest attention'. Gosse continues: 'It was evident that the new comer was embarrassed' and stood 'glancing around him with a sort of shy defiance'. The writer thought he looked like a jaguar or a puma or some such trapped wild animal. He guessed immediately that this strange old man was Hudson, 'the man in the whole of England I most wished to meet'.

Gosse was not deceived by first impressions:

He was remarkably tall, thin and still active, although he was
an old man, his agile movements were those of some graceful wild animal and I felt that to see him at his best I should
observe him on the open pampa of Argentina rather than in
the cramped confines of a committee room in the Middlesex
County Court.

He added: 'his white hair was frizzled and wiry, his eyes dark and
piercing beneath prominent brows. Noticeable were his long ner-
vous fingers of his hands.'

They were introduced. But Hudson was offhand until Gosse
mentioned he lived in the New Forest. Then Hudson warmed up to
reveal that his favourite part of the New Forest was the King's plan-
tation, listening to redshanks. Gosse cited Hudson about the
'thrumming of woodcocks courting'. After the meeting was over,
they turned to books. Hudson said that a second-hand copy of his
Argentine Ornithology had sold for £20 (some £714 today). He was
bringing out a new edition of his *Lost British Birds* (it appeared posthumously) and he presented the first RSPB pamphlet to Gosse for
his collection. They talked about bird protection.

Gosse confided that, before this meeting, he had written to Hudson
in 1915 from the trenches in France where he had been reading
Hampshire Days, which had allowed him to escape the 'squalor and
beastliness of the war'. Hudson had written back. Gosse had mentioned to Hudson that kites should be caught and liberated in Richmond
Park. He considered that this colony of birds would be a 'very pleasing
and appropriate monument to erect in memory of this great champion
of the cause of our rare wild birds'. It's an interesting way of conceiving
of a living monument, but in the end one was built in Hyde Park.

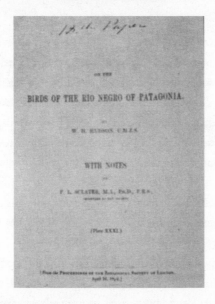

But if the RSPB absorbed Hudson's energy, it was not the only society he attended. A sign of belonging and being accepted is male-only membership of a learned society. Before joining the RSPB, Hudson had already long been granted full Fellowship of the Zoological Society in 1876. Membership was expensive, with a joining fee of £5 and annual subscription of £3. But despite his economic tightness, he retained membership until his death, used the library, attended meetings (for example, in 1895 and again in 1912 to hear his friend Morley Roberts talk) and wrote on the society's headed writing paper.

Not only were Hudson's letters from Quilmes read out aloud and published, but he had three booklets published by the Zoological Society, two of them before he left Argentina. The first came out in March 1872, *On the Birds of the Rio Negro of Patagonia*, with Notes by P. L. Sclater, with the bird species named after Hudson on the cover. The second was on the habits of the vizcacha, also from 1872, and the third was *Notes on the Rails* from 1876 after he had been two years in London (copies are now so rare that I haven't been able to see them). That is, Hudson attended meetings in London while his Argentinean work was still being published. In fact, three letters were read aloud by the chair and recorded in the Minutes of the Scientific Meeting. And he was there. The first was on 2 June 1874, a month after his arrival. On 4 January 1876 the chair read a communication on spoonbills. The last letter was read out on 18 January 1876. It seemed like the start of a career in England. But nothing else by him was published by the society.

He also joined the British Ornithologists' Union in 1893, founded by Prof. Alfred Newton in 1858, who also edited its magazine *Ibis*, which still appears. Hudson wrote one short note for this bird journal. He resigned from the BOU in 1908 because, as his anonymous obituary in *Ibis* stated, 'his nature was exceedingly sensitive and he shrank from causing the death of a single living creature' and thus severed his connections.

Hudson first met fellow Anglo-Argentine Ernest Gibson when he dropped by at his *estancia* Los Yngleses near General Lavalle in Argentina sometime in the late 1860s. Through *Ibis* Hudson continued to read and correspond with Gibson, who published articles on bird life on Cabo San Antonio and around his family *estancia*. His articles appeared from 1879 to his death in 1919. In 1878, the Gibsons owned 65,000 acres of land. Hudson deemed Gibson an 'excellent observer' and cited him in his *Birds of La Plata*. Other members of the Gibson family were members of the society, including Sir Christopher Gibson, and they remain an ornithological family.

Another honour was being elected Honorary Fellow to the Royal Society of Literature in 1912, proposed by the Shakespearean scholar A. C. Bradley. By this date, Hudson's reputation was that of a literary writer. The Royal Society, founded in 1820, had already honoured

Henry James, Thomas Hardy and Edmund Gosse. Hudson was invited to societies like the Society for Psychical Research with F. W. H. Myers in 1893, the Society of Authors with Morley Roberts and the Fabian Society where he heard a 'fine' talk by Anatole France.

He also took his civic duties seriously. In July 1905 he sat on a Grand Jury in Sessions House, Clerkenwell, and examined seventy-nine cases. He could have been excused, and proudly told Roberts he had one case thrown out.

What's clear is that through membership and attending meetings and talks, these various societies offered him a network of London contacts.

Hudson was no bird snob. He didn't follow the twitchers' fashion of stalking rare birds. Any bird would do. In fact, in his poorest days in smoggy London, it was the sparrow that stirred him. His first sight of one was of an exotic bird, never seen in Buenos Aires. Most of the birds and birdsongs that haunted Hudson's years growing up on the pampas were unique to Argentina. He was thrilled, as I've noted, to actually hear his first cuckoo call near Southampton. Once in England, he had to relearn his birds and birdsong. In his chapter on 'Bird Music in South America', he confessed how hard it was to imagine English birdsong from Quilmes: 'I had not heard the nightingale, song-thrush, blackbird, skylark, and all the other members of that famous choir whose melody has been a delight to our race.' Once in England in 1874 'almost every song came to me a surprise', even a sparrow's.

It's said that President Domingo Sarmiento imported sparrows to Buenos Aires as no great city should be without them (Mao would do the opposite and eliminated them). Hudson claimed that he came from a land 'where the English sparrow is not. Now, unhappily, he is there and a great deal too abundant.' Of the 233 La Plata birds, both local and migratory, that Hudson knew personally, one was mistakenly called the 'yellow house sparrow'. Hudson evoked its melody-less song, how it avoided the tree boxes he put up in trees, hounded ovenbirds (*Horneros*) out of their sculpted mud nests and died en masse during the 1867–8 cholera epidemic in Buenos Aires (while he was still there). But the word 'sparrow' confused identities.

In his novel *Ralph Herne*, a doctor comments that perhaps 'that sooty little ruffian, the London sparrow, would have lived it out'. So in 1871 London sparrows hadn't yet arrived in Buenos Aires. However, in a footnote to his 1920 edition of *Birds of La Plata*, Hudson confirms the Sarmiento story: 'Alas! Since this was first written in 1888 the "far-away" relation has invaded Buenos Aires, and as in so many other countries has become a pest', driving the local yellow sparrow into semi-extinction. He confirmed this in a letter to Cunninghame Graham of 7 January 1914, mentioning getting frequent letters from his sister Mary Ellen living in Córdoba, Argentina. She spoke for many when she wrote that the 'hateful English sparrow' hadn't yet reached that area.

For many the house sparrow (*Passer domesticus*) doesn't count as a wild bird as it's so familiar. Yet, despite its Latin '*domesticus*', it hasn't been domesticated. I've watched them fighting and nesting, jumping about my backyard as if all tied together in a bundle. When suddenly they vanished, my cherry tree burst into bloom, and then produced fat cherries. Sparrows had stripped it year after year of tasty blossoms. Heavy wood pigeons try to do the same, but cannot balance on the outer branches. In a 'State of the UK's Birds' report of 2012, sparrows had fallen by twenty million since 1966, due to some virus.

In his *British Birds*, 1897, Hudson noted that the adaptive and hardy house-sparrow is 'sagacious beyond most species', always suspicious and hard to trap. In fact, he wrote a bad poem celebrating them. 'The London Sparrow' appeared in the literary magazine

Merry England in 1883 (edited to lose a line that made the poem jerky, according to Hudson) and was collected posthumously in *Dead Man's Plack, An Old Thorn & Poems* in 1924. He read this sonorous, rhyming poem aloud to his young friend Roberts in 1880.

Hudson, stuck in grim London, missed the 'beautiful world of birds', naming the thrush, the cuckoo and the lark of rural England. But what he really missed were the birds from Argentina, 'nor in this island only: far beyond':

> . . . in hot sweet woods
> the gaudy parrot screams; reedy and vast
> Stretch ibis- and flamingo-haunted marshes

He consoled himself in the 'sad world of London' with his sparrows as sole friends. The sparrow became his 'wingèd Arab of the streets', his 'dusty little scavenger'. Hudson, familiar with rainbow-tinted tanager and scarlet swans, has been reduced to sparrow-friendship, such was his immigrant's isolation. The sparrow was an exile too, nourished on mouldy crumbs. Hudson fed his sparrows from his windowsill (wherever he stayed he fed birds from windowsills). Even then Hudson found sparrows sometimes offensive; birds begrimed with soot, 'chimney-sweep of birds'. He had lost his muse, he said, by coming to London, but the sparrow's pipe 'cheers my exile'. He listened at dawn to his sparrows before 'the muffled thunder of the Underground / Begin to shake the houses' (he lived near the new Metropolitan Underground line). He evoked the hell of London, the sooty slates, the creaking chimney pots, the rank steam of slums, clearly seen from his tower on St Luke's Road. But sparrows bound him to the 'immemorial past' when his area of London was a field. Etymologically, sparrow is Old English, *spearwa*, though nobody seems to know what that means.

As he watched sunrise he thought of Peruvian sun worship and 'the sacred passion of the past' was reawoken. But then urban reality took over and the sparrows appeared, 'prattling fellows' as nature's terminal witness in London. Hudson's desert London is not ours. The air is cleaner, more birds land in my backyard, from jays and rooks to resident robins and thrushes. For Hudson, *one* sparrow is sufficient to banish his overcrowded London and return the city to a silence when human beings no long rule. This poem is a song of despair and loneliness of a double exile, from nature and from his Argentine past. However, Hudson makes do, a trait that David R. Dewar, a Scottish journalist and admirer, calls his sense that 'beauty can be found anywhere', even in a sparrow. There is an Argentine version of the poem, translated by Eduardo González Lanuza, a good poet of the 1920s and friend to Borges. Alicia Jurado found it far better than Hudson's original.

Hudson would identify with humble sparrows, noting *my* sparrows from his turret window. He left crumbs for them. He roamed the London parks, with Ravenscourt Park his favourite, though Richmond had wild spots with woodpeckers and jays, but it was the sparrow that moved him.

He told one last story of a tame sparrow, belonging to an unnamed lady friend, that had survived for eighteen years. This sparrow would let her stroke him, but would only 'allow me to sit by him for an hour, taking no notice, but if I made any advance he would ruffle his plumage and tell me in his unmistakable sparrow-language to keep my distance'.

I'm too impatient to be a birdwatcher like Hudson, who could concentrate silently for a whole day. My bird experiences happen out of the corner of an eye, through a study window. I love seeing a robin strutting about or a blackbird digging for a worm and scattering the earth about, but don't dedicate my spare time to observing them. I'm not sure I even like birdwatchers. They're too obsessive. So what follows is my adaptation of Hudson's vocation.

Jays, when they arrive in my backyard, with white and black crown feathers, and bright blue and black barred wing coverts, are a sudden splash of the tropics. Their call is ugly and loud as they hop about. Hudson cited Richard Jeffries's description of this chattering bird as 'the sound made in tearing a piece of calico'. Hudson called the jay a mocker as the jay, surprisingly, imitates other birds, and can even be made to speak some words in captivity. He found it the 'most beautiful of English birds' and 'the British Bird of paradise'. According to Mark Cocker, one jay plants up to 5,000 acorns to retrieve as food in the winter.

I see in my mind's eye nesting, chatterbox *cotorras* on a huge eucalyptus in the garden at the *estancia* Las Tres Marías. Most people, like the traveller William MacCann, find their din 'disagreeable'. These parakeets are bright green social birds and live in communal nests of woven twigs that hang down from branches. Each couple, Hudson noted, has its own entrance, but they live together. They squeeze into their holes in the giant, disordered nest. They've been a 'National Pest' since 1935 as they gobble crops and are often hunted. I watch a so-called pest-control official stand under the eucalyptus and light the bottom of this nest with a long stick so that it suddenly flares into flames. But the parakeets always return. Hudson once watched them decimate the blossom of a peach tree at his second home of Las Acacias so that a 'pink shower' fell to the ground in what was a bird crime as few peaches then grew. Their din during siesta in baking summers is unbelievable. Hudson described this green parakeet as 'shrill-voiced and exceedingly vociferous'. Both jays and *cotorras* are ugly songsters; their feathers do the singing. *Cotorras* have crossed the Atlantic and reached Germany, and in 2011 were declared an invasive pest in Spain.

Every time Hudson was shown a caged parrot – like the Ranee's caged mackaw – he reverted to the parakeets of La Plata in the dazzling sunlight as a 'corrective' that prevented him hating the caged bird because of the 'imbecility' of its owners. Out free, the parrot 'is to be admired above most birds', he wrote and then added:

I wish I could be where he is living his wild life; that I could have again a swarm of parrots angry at my presence, hovering above my head and deafening me with their outrageous screams. But I cannot go to those beautiful distant places – I must be content with an image and a memory.

I move to another bird that stirred Hudson. He'd noted that pinioned Magellanic geese had been introduced in London parks, though they had little to do with the same geese that wintered every year near his *rancho*. Then he added:

To see them again, as I have seen them, by day and all day long in their thousands, and to listen again by night to their wild cries, I would willingly give up, in exchange, all the invitations to dine which I shall receive, all the novels I shall read, all the plays I shall witness . . . Listening to the birds when, during migration, on a still frost night, they flew low, following the course of some river, flock succeeding flock all night long; or, heard from a herdsman's hut on the pampas, when thousands of the birds encamped for the night on the plain hard by, the effect of their many voices (like that of their appearance when seen flying) was singular, as well as beautiful, on account of the striking contrasts in the various sound they uttered. On clear, frosty nights they are most loquacious, and their voices may be heard by the hour, rising and falling, now few, and now many taking part in the endless confabulation – a talkee-talkee and concert in one; a chatter as of many magpies; the solemn deep, *honk-honk*, the long grave note changing to a shuddering sound; and, most wonderful, the fine silvery whistle of the male, steady or tremulous, the now long and now short, modulated a hundred ways – wilder and more beautiful than the night-cry of the widgeon, brighter than the voice of any shore bird, or any warbler, thrush or wren, or the sound of any wind instrument.

Londoners had no inkling of these strange, wild migrations of honking geese that revived in his mind.

As he wrote that purple passage, he suddenly recalled a very old woman who lived in a mud-built house, thatched with reeds, shaded by old trees by a stream near his own shack. It was a paradise of water fowl. This old woman had a flock of over a thousand geese,

WHISTLING HERON

which, when they saw someone approach, lifted up their necks and burst out in a screaming concert. Hudson: 'I can hear that mighty uproar now!' Then another memory followed, this time of a tale by elder brother Daniel out on his isolated *chacra* near Azul, where a female goose with a broken wing was being urged by her male to fly with him; he would fly up and land again repeatedly, unwilling to give up on his mate.

Another of my own bird incidents reminded me of Hudson. Early one December morning at Las Tres Marías, a strange, tinny whistle was followed by growls. The first thought I had was that nearby sheep were coughing with some lung disease. Then rushing out of the house, we saw two long-necked wading birds balanced on a dead branch of a black acacia. They had tufts or crests and were yellow underneath. One stretched its neck up and let out this whistle, while the other growled loud. Checking, I found that Narosky's bird guide called these birds chiflones or whistling herons.

Chajá by S. Magro.

Hudson in *Birds of La Plata* admitted that he had hardly ever seen this bird and had been unable to observe its habits. So he relied on Félix de Azara who cited its Guaraní name of *Curahí Remimbi*, which means 'sun flute' as they had a 'sweet and melancholic whistle'. It did, indeed, turn its neck into a flute. It augurs bad weather (later that morning a storm hit). Neither Narosky nor Hudson mentioned the strange and ugly musical duet.

Marcos Victoria, a zoologist, noted that Hudson didn't fake knowing this strange bird, a fine example of Hudson's 'probity' as a scientist. On this evidence of how he recorded the whistling heron, we can trust that all that Hudson wrote was directly experienced and when it wasn't, he just said so. He never pretended. Alejandro Di Giacomo in 1988 observed some chiflones in Salto, their eggs, young and nests. He cited Hudson as authority. He claimed they sung like a flutey whistle while flying and sometimes emitted a 'seldom heard' hoarse note. So we had the luck of hearing this seldom-heard call. But he could not explain whether the female fluted and the male growled, or the other way round.

To Hudson, the crested screamer or *chajá* is amongst birds that are like the elephant amongst mammals – due to its swan-like size, its majestic, pale, slate-blue plumage and its domesticity, or love for home. Oddly, it has spurs on its wings. Gauchos often pressed these birds to fight like cocks. Even stranger – Hudson was drawn to eccentricity – it has an emphysematous skin that crackles when pressed. When he first plucked a dead *chajá*, it presented a swollen, bumpy appearance, for just under its skin is a layer of air bubbles. Every time he strolled across Regent's Park, he picked out the *chajá's* singing from its cage. But, he wrote, 'those loud sounds only sadden me. Exile and captivity have taken all joyousness from the noble singer' so that the *chajá* hurries through his 'confused shrieks' as if ashamed. But when a *chajá* soars upwards in its spiralling flight on the acoustic pampas, it releases a 'perpetual rain of jubilant sounds', as intense as any lark. I've often seen and heard these heavy birds.

Hudson complained of its English name, and preferred the onomatopoeic, Guaraní *chajá*. Its scream is its alarm cry. In 1892 he recalled immense flocks around lagoons singing at intervals all night – 'counting the hours' as the gaucho said – in strangely impressive, melodic outbursts. At Gualicho, he stayed overnight in a *rancho* with so many *chajás* singing their nocturnal rush of song that the 'frail rancho seemed to be trembling'. In 1920 he called their song a 'torrent of strangely-controlled sounds – some bassoon-like in their depth and volume, some like drum-beats and others long, clear, and ringing'.

At the Mangrullo *estancia* in the 1860s, on the Western frontier with the Tehuelche Indians, Hudson observed a *chajá* that was the sole survivor of an Indian raid four years earlier. It had wandered the desolate pampas until it found a *rancho* and settled down. It roosted with the poultry and herded the young chicks: 'It was very curious to see this big bird with thirty or forty little animated balls of yellow cotton follow him about.'

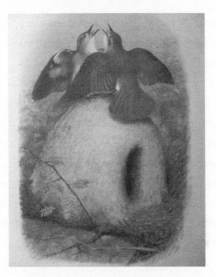

Horneros or ovenbirds on their nest.

Yet, from London, he sensed that these birds were doomed once the land had been cleared of Indians by General Roca and peopled by Italian immigrants, 'pitiless enemies of all bird life'. They would quickly discover that the flesh of a *chajá* is as good as wild goose. They are edible. But Hudson was wrong. In fact, *chajás* are a pest as they land on soya fields and devour the beans. Hudson declared them the 'loudest animal sound of the pampas'.

However, Marcos Sastre (1808–87), a self-taught naturalist and well-known bookseller at the heart of the resistance to Rosas, had also picked on the *chajás* as the most unknown and strange of Argentina's birds and thought a law should be passed to protect them. In his bestselling, patriotic *El Tempe argentino*, 1858, he described this heavy bird in detail, and thought it could be trained to defend lambs from eagles and hawks. Hudson never referred to Sastre, but they emerged from the same tradition – Azara, Burmeister – and could have been companions.

Hudson told the story of his first caged bird. Keeping birds like *cardenales* in cages was and is very common in Buenos Aires. There's still a caged bird market in the city. One undated day, he was strolling in the West End of London when he heard a high, cheerful bird singing. He was so shocked hearing his childhood caged bird singing in London that he froze in his tracks. The bird had recognized Hudson as a companion in exile. Such was the exile's wakening to forgotten birdsong.

Aged eight, in a divided country where everybody was 'chopping heads off', he visited a Methodist clergyman in Buenos Aires for a stay of six to seven winter weeks. The clergyman's wife was his mother's closest friend, and sometimes she would go there on her own. Her return after long weeks was a 'fiesta' in the family. But this time he accompanied his mother. He was a shy child, playing in this Buenos Aires patio, listening to the singing *cardenal* and petrified of making a noise while the busy pastor worked in his study. They were a block from the river shore. When the clergyman left Buenos Aires to go back home, Hudson was given his caged *cardenal*. Then one day back at Las Acacias, it forced its way out of the cage. Hudson followed its song to locate it, tried to lure it back and succeeded once, only for it to escape again. Then one freezing winter, helping to fumigate the rats in the large ditch around the house, he came across its remains.

There's one last totemic bird that had great importance for Hudson. In 1927 the newspaper *La Razón* held a competition amongst Argentine school children to vote for the national bird. The *hornero* (or ovenbird) easily won, coming before the condor, the *tero* and the *ñandú*. You see their solid, globular mud nests on top of poles, on branches, on beams and huts. Hudson held an 'almost superstitious' affection for this dull-looking, rufous bird. He felt it was what an English ornithologist would first want to know about the River Plate. On 5 April 1869 he offered to send nests of oven birds. Gauchos thought it a religious bird that didn't work on Sundays. Hudson tells a *hornero* story of a neighbour in Buenos Aires whose hut had a mud nest on a beam. One of the *horneros* got caught in a rat trap. Released, it flew up to its nest, crawled in and died. Its mate-for-life called and called, flew off and returned in days with a new mate. They sealed up the mud nest with the dead bird inside and started building a new nest next to it. Hudson had once seen ovenbirds build a second nest on top of the first. He particularly noted how *horneros* would sing every time they met up. They face each other and while one emits loud single notes, the other sings rapid, rhythmical triplets in a joyous, marital duo. Hudson kept a print of a *hornero* on the wall of his rented flat in Penzance.

Hudson had a knack for retaining the song of countless birds in his mind, but finding words to capture their song was a problem. He did not attempt to wander subjectively through culture and imagination like Keats with his nightingale's full-throated ease, but in casual prose conveyed a bird's uniqueness. Hudson saw birdsong as intrinsic to the bird as its anatomy or colouring or beak type. And song is emotional. To convey birdsong is a writer's dilemma. Hudson avoided objective descriptions of birds, written by faceless ornithologists. He also disliked illustrations of birds as a way of bypassing words (although his books were illustrated). Recorded birdsong wasn't available, and anyhow disassociated the bird from its song. The song was part of a bird's outburst of life, whether territorial, mating or contact calls, and nothing could recreate it except just hearing it.

That's not to say he didn't try to capture birdsong in prose. An example is the common bunting's 'little outburst of confused or splintered notes' as 'mentally seen as a handful of clear water thrown up and breaking into sparkling drops in the sunlight'. Or

Hudson evoked the sound of the common brown carrion-hawk – you can hear *chimangos* all day long in any pampas wood – as 'rattling castanet-like notes'. But he avoided trying to recreate the tune of birdsong in words, contrary to Argentinian poet Leopoldo Lugones's imitation of a *chingolo*'s song as '*su curí . . . curí qui quío*'.

Chapter 10

The Bilingual Muddle and Literary Style

Hudson's appeal as a writer has never been straightforward. His success was not born of a simple relationship between a writer pursuing a single art form and his single audience. His literary career is a tangled knot involving poetry and prose, fact and fiction, two different languages and two different countries. The muddle of literary genres and languages made him a modern. I wanted to unpick the knot to explore his attitude to poetry and writing, and which language best expressed him, to see what that would reveal of the elusive Hudson.

Maybe doubts about my own writing of poems and novels might help me into Hudson's world? I had tried to balance personal life with family life, with teaching and writing academic books, but it had left me dissatisfied. Ford Madox Ford had promoted Hudson with his friends, but he didn't quite fit. Was he an accidental modernist? His struggle with form and languages was never resolved, because he had to conform to the house styles of the magazines that paid and he just had to earn money from writing. His deep indecisions kept his texts alive. I had to try to pin this down.

Throughout his life Hudson read reams of poetry indiscriminately, without a thought of it being good or bad. 'Love of poetry has been my main passion', he wrote. It's clear that when he read, he was looking for a way of writing that would enable him to see reality more sharply. He read to understand the natural world. Hudson's jettisoning verse to express his vision in prose was to do with scientific realism. Prose brings his reader closer to reality and gives a sense 'of the thing itself'. Plain words catch the actual and good prose should be transparent, should hide its artifice. In *Birds and Man*, 1915, Hudson reckoned that Cowper was a bad naturalist in verse, confusing crows, rooks and ravens (easy to do), but 'Cowper knew better. His real feeling, and better and wiser thought, is expressed in one of his incomparable letters.' So prose was more precise, more honest, and of course, paid better. But he also sought to express his emotional and spiritual side. He was an early victim of C. P. Snow's two cultures, with prose and poetry corresponding to reason and emotion.

So he began with poems. Over his early years in London, he published three poems, 'The London Sparrow', 1883, 'In the Wilderness', 1884, and 'Gwendoline', 1885, in *Merry England*, an illustrated Catholic monthly, edited by husband and wife team Alice and Wilfrid Meynell. This early metrical work suggests a care for words and rhythm implied in the craft of poetry.

But which language to write them in was still a problem. He knew poems in Spanish by poets Francisco Martínez de la Rosa, Juan Meléndez Valdés and Juan de Mena (all read as a boy). Indeed, 'the first collection I ever read was by the Spanish poet Martinez de la Rosa'. These poets were peninsular Spanish ones, and dated. Above all, he knew many gaucho ballads. But there are no references to local Argentine poetry such as *La Lira Argentina*, published as

Argentina's first anthology in 1824, or Esteban Echeverría's popular *Rimas* (1837), including his poem about a woman kidnapped by the Indians, 'La Cautiva'.

Hudson found Spanish a more natural language, less disassociated from prose and speech. As a native speaker, he found Spanish a 'sublimated emotional language', especially compared to English poetry, which was 'artificial'. In an uncollected article in *The English Review* of 1908, poet Meléndez Valdés's tenderness towards birds in 'El colorín de Filis' was 'better expressed in Spanish poetry than in ours'. He approved of the poem's 'seeming artlessness'. He turned to the same poet again who sang, better than any English poet, of the 'pure delightfulness of life with nature'. The poet's free airy assonants clash with the English clink of rhyme that made Swinburne sound mechanical. His watery, pellucid and careful poems were about trees, flowers and streams and seemed to be outside historical time. He was, like Hudson, a rambler, but also the most considerable poet between Calderón and the twentieth century. It shows Hudson's innate good taste, beyond modes.

After Emily's death, at the end of his life, according to Roberts, Hudson cited lines from Meléndez Valdés: '*Es amargo al final de la vida / caminar triste y solo*' (It's bitter at the end of life / to walk sad and alone) and these lines couldn't be more direct. But these poets wrote in classical Spanish.

He told the poet William Canton in 1887 that he was a lover of poetry and that science killed the imagination. So he had made a choice of not killing his imagination. But he also knew how difficult it would be to establish himself as a poet in English. So he began to write prose as if it was poetry, hearing the inner cadences, choosing *le mot juste*, and making sound intrinsic to his style. In a letter to Wilfrid Blunt in 1910 he admitted that 'a real lover of poetry must, I imagine, be something of a poet himself'. Hudson didn't write free verse. To write with rhyme was hard work, chasing rhymes in a rhyming dictionary. So he set to write his prose as if it were poetry, sweating to find the right tone. Morley Roberts reckoned that he only wrote 200 words a day, with much self-censoring. Cunninghame Graham glimpsed pages of the *Far Away and Long Ago* manuscript: 'scored and rescored, they looked like an etching by Muirhead Bone of some great building in construction, with its tiers of scaffolding'.

Hudson had witnessed gaucho poets reciting, not concerned with voice or tone, that is, with aesthetics or beauty. He honed his prose to sound as natural as their ballads. A critic in 1909 defined Hudson's naturalness as a style 'unspoilt by any literary tradition'. He had sensed that his writing didn't fit into the English tradition. He couldn't know that Hudson was reworking the voice behind natural-sounding gaucho poems into a casual, chatty prose. This critic found him to be England's most 'valuable figure in the worlds of letters of today' as a 'natural' writer not a naturalist. That play with natural/naturalist hits the nail on the head. It was a generalized view that you couldn't categorize Hudson. Ford Madox Ford summarized: 'You would, as I have said, think he had never read a book in his life'. And in a way he hadn't; he'd been listening to gaucho minstrels in Spanish. The end result of this confusion about origins and genre was Walter de la Mare's insight that 'Hudson was a poet who merely preferred to express himself in prose'.

A more urgent reason to discard verse and write prose was, of course, financial. How could he have lived off five poems and a novel (that hardly sold) published over ten years? Hudson called his later journalistic writing 'bread and cheese prose' and confirmed the 1881 census where he'd defined himself as a journalist. All Hudson's writing is based on financial need, 'one who had to make a living by writing', he wrote in 1919, but it retained an awareness of something more. Hudson felt that two minds coexisted inside him. One was the plodding 'walking in boots' mind of everyday life and necessity. The other was elusive and unpredictable, the 'sparrow-hawk mind', with its 'swift, flickering glimpses'. He was forced to write 'stodgy stuff' to live, but also hunted for 'prey', like a sparrowhawk. The sparrow-hawk mind works in sudden, surprising images. This was 'seeing something for the first time', a shock of recognition that justified his rambles and his writing. Surprise, after all, is a modernistic trait.

In a 1913 letter to Edward Garnett, he commented that Edward Thomas was 'essentially a poet' who had strayed into prose, like he himself. In a letter to Cunninghame Graham's mother he confessed that he had given up

> what I most valued above everything, the desire to express myself in verse. I could never satisfy myself that I would ever be able to master that delicate, and difficult instrument and so destroyed it. That is to say, I destroyed what I had done and set myself to overcome the wish. Nevertheless the belief remains fixed in my mind that our deepest emotions and the best in us cannot be expressed in any other way.

He stopped writing poetry, but not reading it.

The best defence of poetry is a letter Hudson wrote on 5 August 1911 to Georgian poet Lascelles Abercrombie, friend to Edward Thomas and Robert Frost. There he praised Abercrombie's gift of offering the reader 'a memory of past scenes and experiences vivid as reality itself'. That's a Borgesian insight. It's the *reader's* memory that vivifies a written text. The same applies to Hudson's prose. He awoke a reader's memory. Hudson was aware that Abercrombie might not have witnessed, or felt or smelt that 'road in the desert . . . mapped out with chalk of bones', but it struck Hudson as a reader as luminously real.

Ex-President and General Bartolomé Mitre.

He cited another example: 'the hunger and the heat with blown dried dung in the nostrils eating into the very marrow like a jab of "wild vinegar" that I can experience all over again in your description'. It was Hudson who suddenly felt that pampas smell of dried dung, albeit through the words 'wild vinegar', as it stung his nostrils. Poetry works not because of the poet's experience but because of a reader's associative response. Hudson adds an example from Byron who had never seen a herd of wild horses, yet was able to 'picture them as no one else has ever done'. Words in poems work that miracle of waking *subjective* dormant sensations and memories like wild horses on the plains.

It was the English reader who made sense of Hudson's prose. That's why his prose touched so many who knew nothing of the pampas. He did not integrate prose and poetry so much as bury one in the other; as Massingham put it, his prose is poetry in all but rhyme and metre.

How bilingual was Hudson? In his early letters, his English

spelling was often corrected by Dr Sclater. But he learnt quickly to spell once in England. He never wrote in Spanish. No one could say he didn't write English well. He was chosen in 1948 to be a part of the grand English tradition in the anthology *A Country Zodiac*. And then there's speaking. In a letter of 1890 Hudson confessed that he hadn't spoken Spanish for fifteen years. That is, since he left Argentina for good in 1874. So he was a Spanish-speaker for the thirty-two years he lived out there. The more rusty his spoken Spanish became, the more he felt English expressed his true self. Despite being known in Argentina as Guillermo Enrique, he is only read in translation. Yet, his siblings wrote to him in Spanish and he appreciated poetry written in Spanish.

From letters, I learned that he read the newspaper *La Nación* and historical works by its editor, ex-President General Bartolomé Mitre. He would often quote Spanish proverbs. But by 1904, he confessed to A. R. Wallace that he read little because it was so hard to find books in Spanish 'here'. Yet in a letter to Roberts of 27 July 1905 Hudson mentioned enjoying reading the literary magazine *Helios*. Run by the later Nobel-winning poet Juan Ramón Jiménez for eleven numbers from 1903 to 1904, it published Rubén Darío, Antonio Machado, Unamuno and more. Hudson was in touch with the latest writing from Spain (in fact, Unamuno wrote on Hudson). That same year, he dedicated *El Ombú* in Spanish to Cunninghame Graham, but apart from the odd scattered Spanish words, I'm sure he didn't speak in Spanish. Spoken Spanish became more and more a passive language.

However, he knew countless ballads in Spanish. Barboza, his tough neighbour at Las Acacias, composed his own songs and sang them inexpressibly harshly, like a crow (Hudson's comparisons tend to come from the avian world). Barboza recited strange adventures that illustrated his life philosophy. Hudson said he could 'only recall a few lines' and gave a sample, without the raucous voice, '*En el año mil ochocientos cuarenta / Cuando citaron a todos los enrolados.*' His rhyming translation: 'Eighteen hundred and forty was the year / When all the enrolled were cited to appear.' These lines remained lodged in his memory. He recalled another sample: '*Seis muertos he hecho y cinco son once*', and this time paraphrased in rhyme: 'Six men had I sent to hades or heaven, / Then added five more to make them eleven.' In fact, every gaucho could recite, as having a good or bad voice didn't matter. We know that Hudson recited Spanish ballads around these fires.

Once, a remote Cornish farmhouse reminded him of home, where everyone, dogs and cats included, 'lived in the big smoke-blackened kitchen'. Fuel back home was dried stalks of the cardoon thistle, with dried cow-dung for peat and the 'greasy, strong-smelling bones of dead horses, cows and sheep'. So similar were the kitchens that Hudson almost called the dogs '*pechito*' and said '*mees-mees*' to the cats, 'in gaucho lingo'.

However, gaucho Spanish and English seemed separated in his brain and he did not really believe in translating one into the other. Maybe he knew the two languages too well. To be a good translator, you need to know your mother tongue best. He argued, in a letter to Cunninghame Graham on 6 December 1904, that reading Cervantes's *Quixote* in English was dull. That Spanish literature had to be read in Spanish, because you cannot savour its spirit without knowing the tongue or the people. Following up this insight, after Hudson's death, Cunninghame Graham asserted that you could only grasp Hudson if you knew Spanish, that Hudson wrote English as if he was writing in Spanish. The opposite of bilingual Jorge Luis Borges, who spoke English with his father, but wrote in Spanish. They're part of a notable bilingual tradition that includes Conrad, Nabokov and Beckett. An Argentine critic Roy Bartholomew in his anthology of River Plate poetry included Hudson's English poems

Juan Manuel Blanes, *El Matrero*, c1876.

as an appendix, with only 'The London Sparrow' in a bilingual version. Lombán viewed all Hudson's work as one long translation from the Spanish.

When he was about ten years old Hudson listened to a young Spanish gentleman from Spain. He asked to spend the night at Las Acacias on his slow way south, with only one horse. Hudson remembered two things, his guitar music, composed by Sarasate, and his 'pure Castilian' Spanish, which was music to his ears. The mention of Pablo Sarasate (1844–1908) is odd. This Spanish virtuoso violinist and composer would have only been around twelve years old when Hudson says he heard his work at Las Acacias.

This story prompts another enigma: what Spanish did Hudson speak? Was it the River Plate *seseo*, distorted from Andalucian Spanish and merging the 's', 'c' and 'z', or was it the imperial Spanish from Castile with its lisping *ceceo*? Or was it gaucho lingo? Cunninghame Graham wrote in Spanish to Dr Pozzo in 1934:

'He was Argentine. His slow way of talking and pampas accent always made me think that I had before me a gaucho of old.'

Jorge Luis Borges told an interviewer that:

Hudson's Spanish was very defective. He knew the Spanish used to order a farm worker about . . . but no more. On the other hand, Cunninghame Graham did indeed know Spanish well. Hudson didn't. You see that in the Christian names he used. He got them wrong, gave them impossible names. Well, of course, he worked from memory and memory is usually, at times, too inventive.

And his accusation is correct: Hudson never studied Spanish literature and spoke like a *gaucho*. But it's also incorrect about memory as Borges ignored that Hudson wrote from copious notes.

Jorge Keen, an Anglo-Argentine son of a friend to the Hudsons, left a typescript of a conversation he had with Hudson in London, kept in the Museo Hudson. He first met Hudson in 1903 in the Keen house in Hyde Park Gate, London. Keen noted that Hudson spoke 'gaucho Spanish', that is in short bursts, cutting off the ends of words. He asked Hudson why he had chosen to write in English. Hudson replied:

You see. I loved the Gauchos. As a boy I looked up to them as heroes and used to hang around with them all day long, fascinated by their horsemanship, their dexterity . . . At the age of fifteen or sixteen I had become an expert horseman myself and accompanied them on endless marches and would sit around the fire, sipping *mate*, listening to their tales and sad ballads . . . I choose English to chronicle the gaucho life. In Spanish it would have been like taking coals to Newcastle.

In Keen's recalled dialogue, Hudson repeated that he could recite many ancient Spanish romances or ballads by heart.

An anecdote in 1908 has Hudson receiving a visit from his younger sister Mary Ellen's daughter Laura and her Japanese husband on their honeymoon. They stayed in Morley Hotel on Trafalgar Square and were the first and last family visit Hudson ever had. They spent five days sightseeing and shopping in London. We glimpse another more humble Hudson, whose eyes filled with tears when Argentina was mentioned, who had forgotten the taste of *mate* and who was the same as his photo, except 'sadder and more aged'. Laura Denholm de Shinya tried to talk in English to please Emily, but Hudson insisted on speaking in Spanish. Both visitors were impressed by his prodigious memory. He recited the Luis Domínguez poem placed as the epigraph to *El Ombú* (and then removed).

But perhaps the best image for Hudson's dormant Spanish is an old parrot he tried to befriend in a pub in Hindon, Wiltshire, where he often stayed. Picked up by a sailor in Santa Cruz, Mexico, Polly nipped Hudson if he tried to stroke her. Then he decided to speak to her in Spanish, in a caressing falsetto, calling her 'Lorito' as did the women of the 'green continent'. Polly was instantly attentive and listened, but spoke no word in Spanish. Just low, inarticulate sounds. But Hudson had awoken vague memories of a vanished time. He and Polly became friends at once. This language without words, close to baby or lovers' talk, is how Rima spoke to the narrator in *Green Mansions* in 'inarticulate sounds, affecting me like a tender spiritual music – a language without words, suggesting more than words to the soul'. It was also his mother's manner.

Argentine biographers have argued about how well he spoke Spanish. Alicia Jurado claimed hardly at all, while Ezequiel Martínez

Estrada deemed him thoroughly Argentine. His parents learned Spanish late and badly and mixed with English-speakers. The children, like all first generation immigrants, spoke Spanish like natives. He was a native-speaker. But the emotional Spanish locked in his subconscious, so to speak, kept him a divided being because he could not belong to either the English or Spanish languages.

There's a formal question to deal with. Hudson digressed at whim as a sort of freedom. The rhetorical device of digression moves his text forwards by swerving off the tracks, then returning. His writing and his walking do not follow straight lines. Digression implies going on a journey without an end and getting lost. Anything could enter his 'unplanned' text. In his last book, *A Hind in Richmond Park*, he summarizes his formal approach as an *olla podrida*, a stew into which anything is thrown. That's the best way to describe it, in Spanish.

I once published a small anthology for *PNR*, a literary magazine, of Hudson's prose, guided by an intuition that the prose fragment matched his vision, but that he had needed to write more conventionally as he had to be commercial. So by breaking up his prose into fragments I could restore what made Hudson such an acute poet in prose. I gathered so many fragments that I considered a book with this fragmented Hudson. Then I abandoned the project. It's an apt moment to recover two examples:

> There were always herds of deer on the lands where the cardoon thistle flourished, and it was a delight to come upon them and to see their yellow figures standing among the grey-green cardoon bushes, gazing motionless at us, then turning and rushing away with a whistling cry, and sending out gusts of their powerful musky smell, which the wind sometimes brought to our nostrils.

On the slope of the hill, sixty years from my stand-point were some deep green, dwarf bushes, each bush looking in that still brilliant sunshine as if it had been hewn out of a block of malachite; and on the pale purple solanaceous flowers covering them some humble-bees were feeding. It was the humming of the bees coming distinctly to my ears that first attracted my attention to the bushes.

Chapter 11

Scientific Friends and Foes

Ford Madox Ford recalled how Hudson would often say that he wasn't a literary man, but a field naturalist. He liked belittling his fiction especially when surrounded by novelists. William Rothenstein noticed that Hudson 'affected to disdain his own writing'. He enjoyed the act of writing, but hated rereading his own stuff'. This was not a pose, but a contrarian's position. He always took the opposite view, and that avoids having a fixed view.

Rothenstein further admitted that Hudson was 'really absorbed in literature and cared for good books and liked to discuss them'. It was just that writing fiction seemed less important than being a field naturalist who wrote. He was an outdoor observer of everything he saw, but circumstance obliged him to earn from writing. And that meant being indoors at a writing desk. Fiction was one way of earning and penning nature essays was another, and over the years, he earned twice for each nature piece. Dennis Shrubsall calculated that out of 100 magazine essays only eleven were not reused later as chapters in books.

P. H. T. Harley, an ornithologist, decided in 1964 that Hudson's original observations lay in his South American works, that his British contribution was 'very small' and anecdotal. He had an equivocal attitude to natural selection and was not experimental; in fact, he was a 'scientist manqué'. Hudson didn't published anything on British birds in the *Proceedings of the Zoological Society*. He was a success while he remained in Quilmes. Once in England, and no longer exotic, he found he had to write for the general public. Hudson felt he suffered an imperial prejudice promoted by British naturalists, 'the notion that the homebird is, intrinsically, better worth listening to than the bird abroad'. In a general sense, Borges was right about Hudson, that he sought exile to better understand what he'd lost. A truth lurks behind his witty quip. Hudson needed to exile himself, like James Joyce, in order to write and understand what he'd left behind in his native country. His thirty-two years had given him plenty of matter to spend the rest of his life writing about it. Could my own distance from my Argentinian experiences (I arrived there as a twenty-four-year-old) allow me to write about its literature and culture from the outside better than an insider? I'm not sure, but I do know that writing and exile feed into each other.

The episode over woodpeckers with Charles Darwin (which I've outlined) was crucial in stubbing out his dreams of becoming an ornithologist. Science had moved into the labs and universities. Hudson was left behind, with his acute powers of observation, but no degrees, and no help from his background. The slow snuffing out of his scientific dreams began with his last letter, 'Notes on the Rails of the Argentine Republic', received by Sclater in December 1875, and read aloud and published in the *Proceedings* in January 1876, almost two years after he had arrived in London. As I've mentioned, Sclater tried to help and even co-wrote Hudson's first book on Argentine birds. Morley Roberts considered that Hudson had been

ungracious with Sclater, who had also signed Hudson's naturalization papers in 1900. What I found about Hudson's rancorous
relationships with official scientists is worth telling.

A helpful scientist was Professor Alfred Newton (1829–1907), a
'friend' who assisted Hudson with his paper 'Lost British Birds'. He
was a professor of Comparative Anatomy at Cambridge from 1866
to 1907, a bachelor who publicly defended Charles Darwin. Newton
was a prominent member of the Society for the Protection of Birds
and campaigned against feathers in women's hats in letters to *The
Times* in 1876, which Hudson surely read, knowing that reading *The
Times* was a badge of Englishness. However, there's no mention of
Hudson in A. F. R. Woolaston's life of Prof. Newton.

Despite much common ground in bird protection, Hudson
wrote in 1921 in a letter to Roberts: 'old conservative academic
Newton, a professor of zoology . . . who glared at me, an Argentine,
who dared to come to England and once wrote about birds – the
English birds . . .' He was the 'doyen of the ornithological world',
Hudson wrote in 1921, but he had 'vindictively' attacked Hudson in
an address to the British Association, then, in a letter in the *Times*,
talked of my 'friend Mr Hudson'. Newton combined the worst of
the conservative old squire and the academic professor. Their
unequal relationship was summarized by Hudson as 'the great
Birdist and poor little me'. In a letter of 1893 to Mrs Phillips, Hudson
admitted reading Newton on larks, but that he didn't share his way
of writing, adding, 'I am only a sentimentalist', clearly differentiat-
ing himself from a man Tim Birkhead, a contemporary ornithologist,
calls a 'hard-core museum man'.

Darwin remained a problem all Hudson's life in England. He'd
tried to call on Darwin at Down House, but his son turned him
away. What was going on in Hudson's mind after his woodpecker
tryst? After all, Darwin, besides defending his privacy, was one of the

few Englishmen who had ridden across the Indian-controlled pampas, met Rosas and known La Plata and Uruguay. But somehow Darwin came to embody the scientist-gentleman who kept Hudson out of the scientific world. Whenever he could, Hudson belittled the great man, even after his death.

Nevertheless, the end result was clear: Darwin was right about woodpeckers and natural selection. Hudson was able to 'resist its teaching for years, solely because I could not endure to part with a philosophy of life, if I may so describe it, which could not logically be held, if Darwin was right, and without which life would not be worth living'. Hudson was honest: 'Insensibly and inevitably I had become an evolutionist, albeit never wholly satisfied with natural selection as the only and sufficient explanation of the change of the forms of life.'

He offered a vague Lamarckian view, in which acquired characteristics and oddities are passed on to offspring, according to Roberts because he sought out idiosyncrasies in the natural world. What held his attention was the odd, unpredictable behaviour of animals, not the general rules. He loved direct experience and was anti-intellectual. For example, on the subject of instinctive fear in animals Hudson claimed Darwin's view was 'utterly erroneous'. It was as if Darwin did not pay enough attention while Hudson 'had unrivalled opportunities for studying the habits of young birds'. Hudson knew that science cannot 'explain it all', that Darwinism skimmed over 'unfathomable questions'. We may believe that 'we have it all in Darwin', as he wrote, but there are important exceptions. Nature is not mechanistic.

One dispute with Darwin was over birdsong; it cannot be reduced to male mating competition. For Hudson an excess of beauty made the origins 'mysterious'. Darwin was just 'ridiculous'. Birdsong expressed freedom, emotional outbursts and mad joy. Darwin was accused of being too 'devoted to his theory'. Perhaps Darwin was just tone deaf? When he'd compared the song of the Argentine *calandria* to a sedge warbler, Hudson muttered: 'Darwin's few words were especially remembered and rankled most in my mind.' That phrase 'especially remembered' points to how deeply Darwin's opinions were lodged in his consciousness. H. J. Massingham found Hudson's anti-Darwin asides 'very damaging', but few others did.

Darwin was paternalisitic about Hudson because he had lived unforgettable moments on the pampas. Hudson could have been his son (Darwin's eldest son Francis was born in 1839). I even think that he would have enjoyed Hudson's company as a 'fellow-worker' (Darwin's term). But Hudson's touchiness decided against it.

One example of Darwin's own touchiness concerned Professor St George Mivart, Catholic author of a *Genesis of Species*, 1871. Darwin, so careful not to offend, commented to A. R. Wallace that 'I wrote, of course, to him to say that I would never speak to him again'. Mivart was accused of being a bigot and the sole man 'who has ever, as far as I know, treated me basely'. Now, Hudson cited Mivart approvingly in 1920 as having wisely said in his great anatomical work that there's no such thing as a dead bird. Hudson also endorsed Samuel Butler. According to Darwin, Butler's controversy with him was 'degrading'; Butler 'abused me with almost insane violence'. Darwin's famous public defender, T. H. Huxley, agreed that Butler's fury was 'so ungentleman like as not to deserve an answer'. After reading Huxley's letter of support, Darwin wrote back: 'I feel like a man condemned to be hung who has just got a reprieve.' Hudson always sided with the anti-Darwinists.

In an exchange of letters with Morley Roberts in 1920 Hudson placed Samuel Butler high on his approval list. He had been a sheep farmer in New Zealand. He too had been stirred up by reading Darwin. As an outsider, he attacked the scientific establishment.

Hudson concurred: 'Minimize what Butler did as much as you like, it was he and not Herbert Spencer or anyone else who smashed the Darwin idol and finally compelled the angels of science to creep cautiously.' Hudson was seventy-nine years old and, of course, Butler didn't smash the Darwin idol.

Both Darwin and Hudson reacted to anti-vivisection in the same way. During a Humanitarian League lunch meeting in 1899 in a restaurant in St Martin's Lane, with Henry Salt, Hudson offered to take 5,000 copies of an anti-vivisection paper for the RSPB, and they were given at cost price. While Darwin, in a letter to Prof. Ray Lankester, confessed that vivisection was a subject 'which makes me sick with horror'. Public debate sparked into life in 1874. The anti-vivisection Bill was introduced in 1875, followed by a Royal Commission. It became law in 1876. These dates correspond to Hudson's earliest years in London.

In 1881, in two letters in *The Times*, Darwin preferred the expression 'experimenting on living animals' to vivisection. He outlined his own hatred of cruelty and his role in the Bill, but insisted that science couldn't progress without some experimenting. For Darwin it is 'wanton' cruelty that matters, not a total ban. And this debate continues today in the form of a guerrilla war against laboratories that use monkeys and other live animals. Hudson was more radical than Darwin, and would not have tolerated vivisection in the name of public benefit or the advancement of science.

A. R. Wallace was also an anti-vivisectionist. Hudson befriended the co-discoverer with Darwin of natural selection in 1858, who was a naturalist who'd lived on the Río Negro in Brazil in 1848–52. Unlike Darwin, Wallace was poor, never went to university, was a socialist and feminist, never found the wild indigenous peoples disgusting, and lived rough in the jungles of Brazil and Malaya for over sixteen years. Hudson and Wallace corresponded, but not a word about

Darwin. A letter from Wallace to Hudson on 20 November 1904 attests to the wide-ranging discussion they had. Wallace asked if Hudson had any contacts in Argentina to send him airplants (he didn't), he mentioned George Borrow, asked about adders and asked if he could locate a ghost story by Lope de Vega (he couldn't). Then invited him to visit him at Wimborne, near Bournemouth. Hudson wrote back in an eager-to-please way, beginning with a moan about his gastric troubles. He promised to write to Cunninghame Graham about Lope de Vega, and passed on info about young adders taking refuge in their mother's throat. He referred to a Miss Hopley's *History of Snakes* as she had passed all her notes on to Hudson (he never finally wrote his snake book).

In an undated letter on Zoological Society notepaper, Hudson wrote to Wallace asking him if Rothenstein could paint his portrait as 'one of the leaders of thought of this time'. Wallace had split with Darwin, despite his book *Darwinism*, 1901, over the brain, conscious-ness and spiritualism. So Hudson befriended an anti-Darwinist. In December 1898 Hudson called on Wallace at Parkside, near Bournemouth, and found an old man, dryer than expected. They argued, Hudson wrote to Emily his wife, over bird migration. In a letter to his daughter Violet of 1 December, Wallace referred to a Mr Hudson who'd visited him.

He is very tall, dark and grizzly, something like Dr Spruce but more hawk-faced, very quiet, but pleasant & free in speech. We got on well directly, and talked about his books, and birds, snakes, spiders etc. His is a strong antivaccinationist since he came to England, and has seen several bad results on children of his friends. He appears to be badly off and writes for a living, & has written several novels and tales under another name. We talked a good deal about migration as to which he is rather shaky, believing in birds having a 'sense of direction', travelling accurately N. & S. etc

In another letter to his daughter on 19 December 1898, he repeated this verbal portrait, adding that Mr Hudson was 'handsome' and that 'we got on very well'. He again guessed that he was 'very poor & lives by his writings.' He then added that Hudson 'was cheated by his publisher' and hardly got any royalties from his *Naturalist in La Plata* (it sold 1,750 copies in the first three years, but Tomalin noted that royalties allowed him to travel the shires).

When Wallace wrote his autobiography he only mentioned reading Hudson, but they had a lot in common. They shared being born poor, being self-educated, antipathy to the English class system, experiences of living abroad for long stretches of time, and above all being outsiders to mainstream science. In fact, Wallace's critique of progress echoed that of Hudson. In a 1908 speech at the Linnaean Society that Hudson could have attended, Wallace said:

Our mastery over the forces of nature has . . . brought . . . such an amount of poverty and crime, and [has] fostered the growth of so much sordid feeling . . . that it may well be the mental and moral status of our population has on the average been lowered. If we continue to devote our chief energies to further extending our commerce and our wealth, the evils may increase to such gigantic dimensions as to be beyond our power to alleviate.

Hudson's Abel from *Green Mansions* echoed the sentiment that he loved little children, wild creatures and nature, and 'whatsoever was furthest removed from the common material interests and concerns of a purely commercial community'. What excited other men

– politics, sport, and the price of crystals – were outside his thoughts. Even his manner with women did not arouse husbands' jealousy.

Not hard to imagine, then, Hudson's scientific pride, when he read in the prestigious magazine *Nature* a review of 14 April 1892 by the co-discoverer of natural selection. *The Naturalist in La Plata*, Wallace wrote, was unique among books on natural history, written by someone born abroad and familiar with the wildlife. He praised the clear and delightful style, but most crucially approved of Hudson's observations on the pure pleasure of birdsong or bird dance, a 'wild aerial motion', that mocked Darwin's sexual selection theory of dance. Wallace ended this review of 1892 with 'never has the present writer derived so much pleasure and instruction from a book'.

The next year he reviewed *Idle Days in Patagonia*, but not so glowingly, though finding it 'full of suggestive observations and reflections'. When Wilfrid Blunt visited Wallace in July 1901 it was to meet the grand Old Man of Science, a national 'treasure', enjoying his socialist talk. Despite Wallace's socialism and spiritualism, Hudson had been appreciated.

However, an Argentine way of assessing Hudson's contribution to 'science' came from Marcos Victoria, an emeritus biology professor from the University of Buenos Aires. 'Hudson is our Fabre', he wrote, citing Victor Hugo's famous phrase about scientists as '*savants bêtes*'. He called *Birds of La Plata* the 'masterpiece' of Argentine zoology, following scientific predecessors like Azara, Darwin, D'Orbigny, Bonpland, Burmeister and Muñiz. He saw Hudson in a way that the career English scientists couldn't: that he was a naturalist who studied 'living beings' in their habitats, which is how Hudson saw himself. In fact, Hudson's *Birds of La Plata* is on every Argentine ornithologist's booklist as serious ornithology. It's interesting that translation problems matter less with scientific works than with fiction.

In his lifelong quibbling with official science in the guise of Darwin and the Darwinist establishment, Hudson created his own 'science', very close to ecology, where the personal experience of the observer becomes part of his observation, breaking with scientific objectivity. His credo, prompted by snakes:

> To know the creature, undivested of life or liberty or of anything belonging to it, it must be seen with an atmosphere, in the midst of the nature in which it harmoniously moves and has its being, and the image it cast on the observer's retina and mind must be identical with its image in the eye and mind of the other wild creatures that share the earth with it.

Charles Elton, one of the founders of scientific ecology, wrote in 1927 in his *Animal Ecology* that he had 'learnt a far greater number of interesting and invaluable ecological facts . . . from gamekeepers, private naturalists and from the writings of men like W. H. Hudson than from trained zoologists'.

No lab could replace direct, patient observation. Hudson had been vindicated. But in the end, as his obituary in *Ibis* noted, 'he could never be reckoned among the ranks of the scientific ornithologists'; rather he would endure as a master of the finest English prose.

Chapter 12

Literary Friends and Acquaintances

How did his contemporaries see Hudson? Did knowing him give greater insights into both the man and his work? What emerges is a cubist portrait of some of his traits. I turn to his first literary friend in London, Morley Roberts (1857–1942), who probably knew him best. They shared the struggle to survive as independent writers and a common interest in science and travel.

Hudson and Roberts met in 1880 and became close friends. He was the person who helped prepare the body for Hudson's funeral. He wrote the first biography and published the letters of the man he called 'Hudson', as only Emily called him 'William'. Roberts enjoyed the banter of ideas. Sixteen years separated them, though Roberts considered Hudson as a wise elder brother. Roberts was also very tall and gaunt, with a red beard and, in old age, claimed Margaret Storm Jameson, looked like an 'old sea-captain'. She described him as subtle, restless, self-tormented, humorous, with a powerful intellect. He became a hack writer to support his family,

Morley Roberts.

publishing some eighty-three books, including twenty novels and an evolutionary study of cancer. He was a careless writer, who hardly revised and wrote in bouts of intense work, then relaxed for months.

I read, for example, *On the Earthquake Line: Minor Adventures in Central America*, 1924, illustrated by Roberts himself with six paintings. It was his 'school of crude politics in action' and a very personal book. Given the differences, in essence, very Hudson-like. At the end Roberts thanked the 'vivid talk 'of 'our only hidalgo, who knows his Spanish and his Spaniard, whether a Castilian or the poorest gaucho or mestizo of the pampa' and suddenly I realized that it's the unnamed Hudson. Roberts lived in the outback in Australia on a sheep farm for two years and later three years in Texas in the 1870s and 1880s, so knew first-hand about cattle and sheep farming. His 'utter lack of vanity or self interest' (Jameson's words) drew Hudson out of seclusion into his most intimate friendship.

Roberts introduced Hudson to the illustrator A. D. McCormick. Hudson was McCormick's model for an illustration of a murderous

Morley Roberts (on the left) and W. H. Hudson.

sculptor in Roberts's novel *The Earth-mother*, 1896, though I couldn't spot any resemblance. Roberts, strong-headed but not a murderer, had run off with his already married wife. In a letter to Mrs Phillips, 1890, Hudson likened Roberts to himself, 'poor and Bohemian and literary'.

Between 1880 and 1884 Roberts was a weekly visitor at the boarding house on 11 Leinster Square for the Wednesday open house. He took to Emily, found her humble, devoted, ageless and with glorious red hair. She was a soprano, though of the 'second class', who gave twice-weekly singing lessons to Roberts, himself a chorister at school. Emily would play the piano, Roberts would sing arias from operas and Hudson would sprawl on a sofa, cackle aloud and listen. Many evenings were spent this way. Hudson also enjoyed song, but, curiously, didn't sing himself. Maybe Smith's deep shame in trying to sing in *A Crystal Age* is based on his own experience: 'But when I remembered my own brutal bull of Bashan performance, my face, there in the dark, was on fire with shame; and I cursed the ignorant, presumptuous folly I had been guilty of in roaring out that abominable ballad . . .' Roberts dined regularly and enjoyed reminiscing about his outdoors life abroad.

Roberts kept notes about these meetings, like a minor Boswell, and copied some down in his biography. Here's one, dated 29 December (but no year):

Hudson showed me and N. his queer, fantastically queer, hard side to-day. He brought out a manuscript page of contents for one of his new books. In this one paper was called *The Vanishing Curtsey*. N. asked if that meant the going out of the custom of curtseying. He didn't hear what she said and she repeated it. Curiously enough he answered her almost rudely. Perhaps he was annoyed to be reminded that he was now at times a little deaf.

H. 'Why, whatever else could it be about?'

As a matter of fact she and I had both fancied it might have some reference to some mystical story, as he had been talking of spooks. However, that was not his hard side. Among his papers was an inconsiderable one he proposed to destroy.

H. 'I've cut it out and shan't print it.'

R. 'Why not? Isn't it good?'

H. 'Not particularly. I did a sketch of a man, and to make it interesting I exaggerated his good points and published it in a magazine. And this fellow actually wrote to the editor saying he was the man described and asking who I was!'

N. 'Did you tell him?'

H. 'No! Certainly not. Why should I? He had no right to do it.'

R. 'He oughtn't to have done it, of course.'

H. 'It annoyed me very much.'

R. 'After all, Hudson, it was only harmless vanity on the fellow's part. I don't see why you shouldn't reprint it.'

H (*pettishly*). 'Well, I won't. I'm not going to gratify his vanity. He shouldn't have written.'

N. 'But, Mr Hudson, perhaps he only just wanted to get in touch with you again.'

H (*indignantly*). 'Yes, that's just it, and I didn't want him to get in touch with me. I've had all I want out of him and he doesn't interest me any more . . .'

In September 1893, Roberts was sketching with watercolours on Shoreham shingle beach with Hudson when three young girls dipped into roughish sea to swim and were carried out by the tide. They screamed for help, so Roberts rushed into the waves and saved one, then a second. Hudson couldn't swim, so pulled the girls on to land, helped by a vicar, Revd F. N. Harvey, who'd arrived at the scene. The third girl grabbed Roberts and dragged him 15 feet underwater, so Hudson strode in and pulled them both ashore. Roberts, exhausted, lost consciousness, while Hudson was soaked through. Roberts thought he'd lost his pipe, as he'd rushed into the water still smoking it, but Hudson had eased it out of his mouth. Saving three young girls, one of them local painter James Aumonier's daughter, got them into the newspapers.

Hudson relived the same incident in a letter to Emily, his wife, on 3 September 1890, adding that he went in up to his chin to rescue the girls while Roberts swam out again. A copy of a certificate issued by the Royal Humane Society, patron the Queen, and dated 3 September 1890 is held at the Museo Hudson. It praised Hudson 'gallantly' saving three young girls.

Roberts and Edward Garnett both noticed that Hudson would scoop up saltwater every time he went down to the sea and drink it, but it contradicts his odd claim that the 'sea is not nature – to have it before me is a weariness'. A contemporary who cupped seawater to drink it was Edmund Gosse and he sought its mild, irrational power. Gargling in salt water was a restorative.

Roberts confirmed Hudson's quirks. How he hated dates and obscured his date of birth, and loathed biographies. How he refused a bath at his North Kensington home and kept to the zinc tub, with hot water brought up to him. He went into details about his ill-health, naming doctors. How he wrote to his wife Emily every day that they were separated. He informed us which towns Hudson preferred like Wells-next-the-Sea or why he concealed the village of Martin in Wiltshire under the invented Winterbourne Bishop. But most of all, Roberts and Hudson traded views over science. He decided that Hudson was a Lamarckian and introduced him to scientists like Sir Arthur Keith. Even Ezra Pound met him through Roberts, though Pound talked too much, according to Hudson. Pound put him into *The Cantos* (LXXIV) as 'Huddy going out and taller than anyone present'.

In a letter Hudson remarked that Roberts had succeeded 'in astonishing the leading scientists of London with his work' on cancer in *Warfare in the Body*, which had been his hobby over many years. Hudson regretted his devotion to writing novels for a living when his genius was for popular science. Roberts and Hudson craved scientific recognition, but both strayed into writing fiction. Over forty-two years of friendship, Roberts admitted that Hudson hid his motives, and that he 'spoke intimately of women very rarely' and hated pornographic humour.

Another friend, met through Roberts, was George Gissing (1857–1903). In a letter to William Rothenstein of 22 December 1903 Hudson wrote, 'I was one of Gissing's half dozen closest friends', and many letters between them attest to that. Roberts had been at

Owens College, Manchester, with George Gissing. Hudson was aware of the 'ever-lasting half-whispered sort of tittle-tattle going on about poor Gissing', and it alerted him to prying future biographers. Gissing had been expelled from university for stealing from friends to support a prostitute he later married. He travelled a year in the United States and came to London in 1877 to write. He later moved to Italy, then France to live with his French lover (they couldn't marry) Gabrielle Fleury, and died at Saint-Jean-de-Luz.

Hudson read his pessimistic novel *Born in Exile*, 1892, three times. Morley Roberts thought it Gissing's 'greatest book'. The themes of class, poverty and exile touched Hudson. But there's more. He'd iden- tified with the protagonist Godwin Peak, who called himself a lodger and an alien in London, without class privileges. Peak was shamed he wasn't born a gentleman, had to hide his northern accent, even becoming a hypocrite in wooing a woman above his trader's class by pretending to become a priest. Several discussions follow about the clash between the new science of Darwinist materialism and Church traditions. As an intellectual rebel he'd written an anonymous paper mocking traditional beliefs, but was forced to accept a modest post as a scientist outside London. He died in Italy, an outcast, despite inher- iting some means. One sentence conveys why Hudson, the exiled foreigner, read this book three times: 'We talk of class distinctions more than of anything else – talk and think of them incessantly.' From the letters between Gissing and Hudson, it's obvious too they didn't see eye-to-eye about how being in nature calmed the mind.

They met on 24 March 1889 at the portrait painter Alfred Hartley's Chelsea studio. Gissing noted in his laconic diary: 'very striking face, gentle, sympathetic manner'. On Saturday 5 October, Gissing met Roberts for dinner. He'd baptized their little group the Quadrilateral. They drank whisky with 'a vast amount of talk'. A week later Gissing was invited round to the Hudsons' Wednesday

Joseph Conrad.

open salon. Gissing's first impression of Emily: 'the poor fellow is married to an old and very ugly wife, who formerly kept a boarding house'. They argued in the 'usual pleasant evening of talk' over Hardy and Dickens (Hudson admitted: he 'exasperated me'), next met at Saville's club and exchanged books. When Gissing moved abroad, he wrote long letters about Greece and 'suffered homesickness' when he read Hudson's *Hampshire Days*. Hudson sent him a dictionary to read the *Quijote* in Spanish. Gissing, close to Conrad, was surprised that Hudson was not an Englishman, as he'd found Henry James too 'déraciné'. He 'greatly rejoiced' over Hudson's government pension in 1902.

After Gissing's death, Morley Roberts wrote a fictionalized biography, *The Private Life of Henry Maitland*, 1912, that shocked Gissing's widow so much she refused to acknowledge him in the street. He'd attempted to be absolutely sincere and pretended that it had been dictated by a J. H. and merely revised and edited by himself. Hudson doesn't appear, but it did reveal all Gissing's faults from stealing and

Ford Madox Ford.

being expelled at his college to his relationship with his wives, a warts-and-all story. Hudson read his friend's biography with 'intense and painful interest'. He felt 'very bad' and 'make me wish you [Roberts] had never attempted this thing'. Gissing's family 'will hate and curse you for it'. It was a lesson in biographic intrusion for Hudson. Roberts would go on to write a very different biography of Hudson after his death.

Hudson also befriended his novelist brother Algernon Gissing and helped him get a grant from the Royal Literary Fund. He and Emily read his works. Through Roberts and the Gissing brothers, Hudson was made to feel at home in literary London.

On 26 January 1906, Hudson was introduced to Joseph Conrad, another alien, by the maestro behind Hudson's modernistic reputation, Ford Madox Ford. Ford had met Conrad in 1898 and became his collaborator in writing fiction like *Romance*. According to Conrad, he and Hudson met some ten times. Conrad, while reading *Green Mansions*, looked up and famously told Ford Madox Ford that

Hudson's 'writing was like grass that the good God made to grow and when it was there you cannot tell how it came'. This is one of the great clichés about Hudson. What did Conrad mean by that? Something natural, like a voice or a personality that shines through the prose? Conrad thought that Hudson's secret as a man and as a writer was 'impenetrable' and 'uncanny' and, in a letter of 1923, evoked his 'fascinating mysteriousness'.

Virginia Woolf, in an anonymous review in 1918 – my epigraph – of *Far Away and Long Ago* (I still remember my excitement at discovering her identity) hit the nail on the head hailing her discovery of 'the whole and complete person whom we meet rarely enough in life or in literature'. Through voiced prose Hudson had created a sense of a complete person. Conrad also remarked how Hudson's correspondence with Garnett 'reproduces the accent of his talk'. He wrote like he talked.

Ford (called Hueffer by Hudson; he'd change his name in 1919) told the story of Hudson's dropping round on Conrad at Pent Farm, Postling, Kent. A tall man was seen outside stalking around. Conrad thought he'd come about exchanging his mare for some Shetlands or maybe it was a bailiff. Then the man outside, looking like a Spanish mayor, asked in his slow voice to see Conrad. Hudson would pause before he spoke, and then stare at you with a 'sort of humorous anticipation'. He was not at ease with literary men, would often burst out, as already noted, with 'I'm not one of you damned writers: I'm a naturalist from La Plata.'

Then Ford touched on his inability to discover how Hudson got his effects. He assays one explanation, rather like Woolf's, that when you read Hudson, it's as if 'a remotely smiling face looked up at you out of the page and told you things'. And those things remained in your mind as your own experience. Ford then reported a dialogue with Hudson:

'It's the simplicity of your prose,' I would protest. 'It's as if a child wrote with the mind of one extraordinary erudite.'

He would answer: 'You've hit it. I've got the mind of a child. Anyone can write simply. I just sit down and write . . .'

'You know it isn't,' I would protest. 'Look how you sweat over correcting and recorrecting your own writing . . .'

The conversation wound on. For one, art is clarity, economy, surprise. For Hudson, the other, loathed showing off, complicating his prose with style, as he was not a literary artist. Then Ford would throw up his hands in despair, and begin all over again. But we have enough in this dialogue: the reader sees through Hudson's finely tuned prose to a man with a child-like mind, talking of unique experiences, as if you hadn't heard him before.

Ford organized a tribute to Hudson in the *Little Review* of May–June 1920, the first of its kind. He wrote a rambling piece about Hudson being 'master of the English tongue'. His clue was Hudson's selflessness. He approached nature with his whole mind, and wrote in a spirit of tranquillity. He met Hudson in the Riji café in Soho with Garnett, Galsworthy, Wells and Hilaire Belloc, who was boasting of being a Sussex man, born and bred. Belloc said you could ride from Crystal Palace to Beachy Head with only four stations. 'Five,' said Hudson with his deep raven voice. Belloc insisted. Hudson just repeated 'Five.' Belloc named the four stations and Hudson then added in the missing 'East Dean'. Belloc admitted he was a fool to forget East Dean. Hudson sat motionless, grave, unwinking. Belloc once admitted to loving his novel *A Crystal Age*, rereading it several times, without knowing the author was sitting opposite him in Soho.

Ford guessed rightly that Hudson's Argentine birth and 'long racial absence from these islands' saved him from the infection of the 'slippy, silly' way the English handled their language. Ford added that his English was so natural 'you would think he had never read a book in his life'. This is an apposite insight for Hudson wrote out of an alien tradition. You recognize Hudson readers for when asked about him their face lights up, become animated, eyes alive. Hudson could 'teach' us all.

Here was success for Hudson, not of sales, but of a modernistic reputation, and all thanks to Ford's selfless boosting. In the same magazine, the *Little Review*, John Rodker (1894–1955), modernist poet and in 1919 publisher of T. S. Eliot and Ezra Pound, argued that only foreigners like Hudson and Conrad could write 'live English', their senses not dulled by 'traditional thought-forms'. Their brains are 'brand-new and respond immediately to life'. Hudson gives his readers a 'solid chunk of life'.

In 1931 Ford returned to Hudson and Conrad (and Crane and Henry James) as a 'group of foreign conspirators plotting against British letters' (H. G. Wells's jibe). After all, Henry James and Conrad spoke to each other in French. Conrad 'loved' Hudson personally, with an 'unbounded' admiration for his books.

To show how well Hudson guarded his personal life, Ford guessed wrongly, that he had arrived in 1882 and was the first member of his family to visit England for over 2,500 years, a typical Ford exaggeration. Hudson, in Ford's description in 1906, was still very lean, *very* tall, slow in motion, with weather-beaten cheeks, small eyes and a hidalgo's beard, and embodied 'gentleness and infinite patience'. When they strolled together, Hudson was mostly silent, occasionally butting in with a word of dissent. Ford is the only witness to describe his accent. Thirty years in England had modified his original American East-Coast accent as 'neither English nor American, but very scrupulous'. But what Ford would not have known is that Hudson's way of talking was a gaucho way.

The pampas imposed a particular mutism and Hudson, like all gauchos, would interrupt the conversation with interjections, with witty phrases or cackles of laughter.

Hudson's persona also granted Ford an insight into writing, for he realized that the 'patience of the field naturalist goes with good prose'. If you are not 'simple', you cannot observe. If you cannot observe, you cannot write, so you must observe with humility, the first characteristic of the great writer. Behind Hudson's prose and field observations lies *self-abandonment* and *patience*. Furthermore, there's self-knowledge. As Hudson noted just before dying: 'My credentials are those of a field naturalist who has observed men: all their actions and their mentality. But chiefly himself, for to know others a man must first know himself.'

However, Hudson never really appreciated Conrad's fiction. In a letter about *Nostromo*, he found 'the S. American atmosphere is false, meaning principally the mental atmosphere – the mind of the natives'. He preferred Conrad's memoirs. Hudson had become the arbiter of English writing about South America. What did Conrad know about South America, except from talk and books? He'd spent three days at La Guaira, the port for Caracas, and twelve hours at Puerto Cabello. Yet, Hudson was wrong. In *Nostromo*, 1904, Conrad *invented* a plausible country, fusing several Latin American countries.

Hudson met the dashing Scottish aristocrat, traveller and politician Robert Cunninghame Graham, 'my impulsive friend', in the Café Royal in London in 1890, before the latter published his first story in 1895. He would, however, always remain an 'amateur' writer, with an aristocrat's scorn for earning money from writing (it tells, unfortunately). Hudson knew this meeting was predestined. In a letter of 10 March 1890 Hudson wrote:

it would be strange indeed if I did not know the Pampa, seeing that I was born there, and as I have the feeling for it which
each one of us has for 'his own, his native land'. It is always
a rare pleasure to meet anyone at this distance with whom
I can compare notes about it.

The quotation within the letter is, not surprisingly, from the
Bible, Isaiah 13:14.

He dedicated his short stories, *El Ombú*, to this late-developing
writer, who had visited and loved Hudson's homeland of Argentina.
This crucial Argentine experience linked them deeply. Of all the men
Hudson met since his arrival in 1874, Cunninghame Graham was the
sole to grasp the reality of Hudson's pampas background and for-
mation. Cunninghame Graham had travelled three times, between
1870 and 1878, in the pampas and in Uruguay and it remained
imprinted in his soul.

A seventeen-year-old Robert Cunninghame Graham.

There's a photo of him in Tschiffely's biography as a seven-teen-year-old dressed as a gaucho with beret, whip, poncho and *bombachas*. He'd experienced gaucho hardships, for example, when he rode 600 miles up the Paraná to Paraguay and the Jesuit missions in 1872, researching his book about the abandoned missions, *A Vanished Arcadia*. In 1878, he also tried to farm near Bahía Blanca on the Indian frontier with a friend and had to escape on the gallop to the nearest fort at Mar del Plata as marauding Indians burnt down his shack.

He wrote extensively, in the form of static sketches, about the Indians, the horses, the plains. He was in love with the 'picturesque' and wild, and loved riding the pampas. He famously defended the 'Red' Indians of the United States and Argentina and found their treatment a 'disgrace'. He'd travelled the whole of Latin America from Mexico to Venezuela, Colombia, Brazil and Paraguay, but thought, with Darwin, 'that perhaps the pampas have charm greater than anything'. In the Quilmes Hudson museum, both writers are remembered.

From the late 1890s to his death, Hudson was at last understood, as few of his few other friends even spoke Castilian. Cunninghame Graham's mother was the daughter of a Spanish woman from Cádiz, who often took him with her there on holidays so that he became bilingual. His wife Gabrielle was Chilean-born, with a Spanish mother, though moved to Paris at the age of twelve. She too was a bohemian and writer, spending hours researching a two-vol-umed life of the mystic Santa Teresa and smoking up to 200 cigarettes a day (which eventually killed her). Or so she said. It emerged in 1985, from family papers, that in fact she was Carrie Horsfall from Yorkshire, who ran away and invented herself. She had been a mistress to some rich South American and learnt Spanish. But she always had kept in touch with her mother. Even her death certificate in 1906 faked who she really was. Hudson never knew her real identity. Cunninghame Graham whiled away hours practising with a lasso while at his estate in Scotland and always wore his gaucho belt, even in formal dress. A farmhand I know owns a gaucho belt, covered with sewn on coins, including a Norwegian kroner and bits of silver. He still wears it as his Sunday best. In fact, all the gaucho accoutrements seen on Cunninghame Graham are still in use.

Exactly why Cunninghame Graham was so deeply bitten by Argentina was related to riding and freedom. He knew the pampas before wire fencing, before meat-freezing and the parcelling up of land into great *estancias*. Among the wild indigenous tribes and tough gauchos, young Cunninghame Graham, nicknamed '*el gringo*', with his Van Dyke beard and tousled red hair, became a centaur. He lived for his *criollo* horses, and saved one from hauling trams in Glasgow, renaming it Pampa. It was a real coincidence. He'd recognized the Eduardo Casey brand on the horse from the 'Curumulan' *estancia* near Tandil. He dedicated a later book, *The Horses of the Conquest*, 1930, to his 'black Argentine – who I rode for twenty years, without a fall'. He even rode him to the Houses of Parliament while a socialist MP. When his horse died, he 'never felt anything more'.

Hudson shared his passion for wild riding, gaucho-style. But Hudson, an instinctive rider, refused to ever ride in Rotten Row, though Cunninghame Graham pranced up and down the sawdust tracks in Hyde Park in his bowler hat. He lacked the latter's vanity. Don Roberto couldn't pass a mirror without preening.

Actually, the two men did not meet in the flesh that much, but corresponded and that one-way correspondence has been collected (of course, Hudson destroyed the letters he received). Hudson would drop round to Cunninghame Graham's mother in Chester Square for tea, breaking his habit of not socializing with the rich. The two men were separated by age, birth and wealth, but joined

Cunninghame Graham on Pampa in Hyde Park.

by one having tasted being a gaucho as a seventeen-year-old and the other just having been born one, for he was, in Cunninghame Graham's words, 'at heart an old-time gaucho of the plains'.

When Hudson died, Cunninghame Graham set up the committee to create the monument in Hyde Park. In a letter in the *Times Literary Supplement* of 21 December 1922 he called Hudson 'a great Englishman'. Years later, in 1936, Cunningham Graham found himself again in Buenos Aires for the last time. He knew he was dying, but decided it would be his last chance to ride with the gauchos. He insisted on visiting Hudson's rediscovered *rancho* Los Veinte-cinco Ombúes, with Dr Pozzo, who had relocated it.

In a letter to Morley Roberts, he evoked the *rancho* as a shrine, with its primitive doors and 'its air of aloofness from everything modern' and added 'little I think has altered', except that only three of the twenty-five ombúes had survived. In the same letter he wrote 'the same flocks of birds, *tijeretas, viuditas, bien-te-veos* and *horneros* still haunt the trees that have grown up in the deserted *chacra*'.

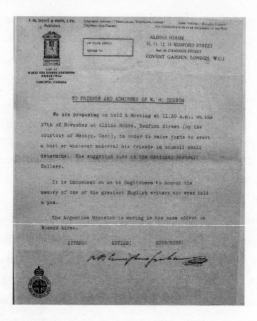

His Spanish was still perfect (he was bilingual) and he'd just finished writing a preface in Spanish to Hudson's autobiography.

Back in Buenos Aires, in the grandest hotel in South America, the Plaza on Plaza San Martín, he suddenly died. His body lay in state for twenty-four hours in the Cervantes theatre, with even the President of Argentina, Agustín Pedro Justo, paying his respects. Then his coffin was shipped home. Rubén Ravera, director of the Museo Hudson, tried to locate the exact room in the Plaza hotel, but records have been lost.

In his way, Cunninghame Graham was a greater promoter of Latin America than Hudson, though he knew his knowledge was superficial. Theodore Roosevelt praised him for what 'you and Hudson have done for South America'. He once honestly described himself as a 'foreigner writing for foreigners of a country foreign alike to reader and writer'. But that wasn't Hudson's position. On

Cunninghame Graham's Romantic tomb on Lake Menteith, in the ruined Augustinian monastery on the island of Inchmahome, you can see a carving of his Argentine cattle brand, because he had actually branded cattle. Branding is a tough job. The bullock is lassoed (from the Spanish '*lazo*') to the ground; the iron brand is heated on a fire, and then you hear the sizzle of flesh as the *estancia* brand is burnt on. Nothing was more intimate to *campo* life than young bull branding and castrating. I saw it done in the early 1970s with large scissors. The testicles were tossed on to the embers to be eaten with steaks.

This aristocratic writer, so close to Hudson in direct experience, was very tall, upright and fierce eyed and with a personality that shone through his writing. Like Hudson, it's the man behind the prose that moves you. Hudson was lucky in this friendship as Cunninghame Graham was a 'cultural go-between', introducing Hudson to Conrad and countless fellow writers. In a letter to Herbert West of 3 August 1929 he said that, unlike his literary friends, he and Hudson talked of serious matters, of horses, Indians like Catriel and gauchos, that they were 'equestrian philosophers'. His intimate name for his friends was 'Don Roberto'.

Poet Coulson Kernahan was not the first to compare Hudson with Cunninghame Graham; both had a 'touch of the Spanish hidalgo' about them – beards, eagle faces, tall. But also because Hudson was so chivalrous that Kernahan murmured when he saw him, 'Don Quixote, surely'. Violet Hunt had also found him 'so foreign-looking'. Kernahan bumped into Hudson at a salon run by the Bostonian poet Louise Chandler Moulton (1835–1908), who summered in London between 1876 and 1898. She found Hudson almost tragically lonely. His sensitive, gaunt and grim face suggested a man who was poor, but she'd glimpsed more, his 'soul-loneliness'. Hudson was ill-at-ease with the rich and his frank and natural habits didn't fit. He told Kernahan that he was a recluse, awkward at such literary salons.

Though they met seven or eight times, he never broke through Hudson's reserve. He hadn't a clue if Hudson was married, didn't even know he was a writer. Hudson actively withheld information, a trick picked up as an exile. He would chat about Spanish literature in his low-pitched, pleasant voice. His talking style was simple, frank and lucid, matching the way he wrote. He shrank from boasting. For Kernahan, his obscurity as a writer derived from his personality.

Hudson was buried in the same cemetery as Richard Jefferies. They never met, but Emily Hudson compared them, as did Samuel Looker. Today, there's a Jefferies Museum in Sea View, Goring, where he died (although he was brought up at Coate Farm, near Swindon, before the M4 sliced into his hills). In September 1899 Hudson stayed in the Jefferies' house in Goring and wrote the opening chapter of *Nature in Downland* there. I assume that Hudson wrote sitting down, unlike Tolstoy or Hemingway. Hudson had read his autobiography *The Story of My Heart*, knew his utopic novel *After London or Wild England*, his novel about the boy Bevis, and works on poaching and gamekeepers. Jefferies was an outdoors sensualist, a sick nature mystic, and like Hudson, survived from his writings.

But Hudson was critical of his fellow-naturalist. He found *The Story of My Heart* a dangerous book, repelled by its lack of reticence, its 'intense unnatural feeling', its 'insanity'. He complained in a letter to Garnett in 1905 of Jefferies's 'deadly want of humour'. He saw his bust in Salisbury Cathedral and disliked its expressionless face more than ever (though I couldn't locate it in the cathedral). He even fell asleep trying to read the novel *Bevis*. But Jefferies was a genuine critic of urban hypocrisy, especially the Victorian lack of passion and sincerity; he longed to purge late-Victorian London and hated 'house-life'. 'Begin wholly afresh. Go straight to the sun', he advised, yearning for direct contact with the earth, beyond culture and the grip of dead tradition. But due to ill health his loathing remained bookish. Nevertheless, Hudson was haunted by his early death aged thirty-nine, by his unfulfilled dreams. Later, Edward Thomas dedicated his biography of Jefferies to Hudson.

Now we come to an acute insight about Hudson. Richard Curle, a Scottish writer who met Conrad in 1912 and became his literary executor, accompanied Hudson to Wells-next-the-Sea in November 1912 to watch the pink-footed grey geese migrating from the north. Hudson had previously invited him to the Mont Blanc restaurant, upstairs, to meet Garnett as he'd enjoyed his book and stories in *Cornhill* so much. Edward Thomas told Curle that he was very privileged as Hudson never asked anybody to accompany him in his nature rambles, as he learned to his cost. Curle hadn't read Hudson, who said 'a day in any wild place with another is to me, a wasted day'. He sensed two sharply contrasted Hudsons. The London Hudson enjoyed company, was curious about people and relished talking. The other Hudson stalked birds in the country, retiring into himself.

Hudson's Wells-next-the-Sea during the autumn and winter was out-of-season, remote, wild and desolate, one of the 'few' places in England he could recommend. Here they walked off to the woods beyond the tidal harbour and the drained marsh. Hudson was mute, on the alert, observing everything. When the geese flew past honking, Curle saw that Hudson's face 'showed nothing', that he was completely absorbed in his listening and 'incomparably happy'. He had no wish to share his thoughts. Curle, however, felt so awkward, so unwanted, that he cut short his stay. A veil had been lifted when he had seen that impassive face as the geese flew overhead. Hudson, 'the least conceited of men', had a 'completely integrated personality'. Migrating geese in Norfolk reminded him of the thrill of migrating geese in Quilmes.

If I explore this insight a bit more, I can clearly discern Hudson's detachment. He'd once told his friend Cunninghame Graham that 'I have a colder mind'. He'd praised his Scottish friend for uniting those two qualities of 'intense individuality and detachment' which enabled him to identify himself even with those most unlike us, with birds and animals. He was talking of himself. At Wells he was truly happy alone, 'listening to the sea-wind in the pines'.

Curle wrote an obituary on Hudson in 1922 and made further striking points. Hudson was a man full of curiosity, without poses, 'extraordinarily' sincere, 'but I have never known a man who could put a more withering contempt into a judgement'. His opinions were completely independent of others, and he never talked for effect, yet his words lingered in the memory. He added: 'I have seen him stand in a wood, alert and silent as a Red Indian and with some-thing of a Red Indian look in his hawk-like features.' Many times Hudson has been identified with the Red Indian; it must have been a way of saying that he was other, totally distinctive. Curle watched Hudson writing *Adventures among Birds* at a table of their joint sitting room in a cottage in Wells by the harbour, with steady concentra-tion every morning. 'Nobody', he wrote, 'ever had a stronger or more self-contained character than Hudson.'

On 22 August 1922 Morley Roberts thanked Edward Garnett for attending Hudson's funeral: 'You really loved him & there were many chattering fools & busybodies who didn't.' He added that 'he had so great a trust in your judgement'. Not a shadow of doubt about Garnett rescuing Hudson's career as a writer from the day when, as a publisher's reader with 'uncanny insight' (Conrad's phrase) at Duckworth's, he pushed to get *El Ombú* published, calling it a 'work of genius'. What he said went. If a story had to be left out, or a passage moved to the back, then Hudson conformed. He had blind faith in his taste.

Garnett edited Hudson's letters to him in 1923 and the first anthology of Hudson's writings in 1924. He organized literary lunches and introduced Hudson to literary people, making him feel at home in London as a writer. He alone could call him 'Huddie'.

They first met in 1901 in the publisher J. M. Dent's offices. Garnett evoked this sixty-year-old Hudson as very tall and thin, with a remarkably small head, hidden in a very long overcoat. He was extraordinarily silent in his movements, with very bright eyes, a hooked nose and 'resembled a bird'. When it came to the books, Garnett, crucially, preferred the Argentine books as his English ones lacked the 'freshness of his early first-hand acquaintance with nature' (I agree). Garnett found many faults with his fictions, from *The Purple Land* as broken and spasmodic to Rima's leaden talk.

Now we come to Garnett's best insight, close to Curle's earlier one, about Hudson being an emotionally and imaginatively *cold* observer. That coldness was an impersonality that gave him the power to encounter strange people and chat with them as well as to note nature's quiddities. His 'casting a cold eye' came from looking at all life, and not taking sides about human beings. On his scale, an ant was as miraculous as a man. Such a wide-ranging spectrum made him appear cold. Hudson's natural response to somebody's ebullience was 'at most a monosyllabic abruptness', another shrewd insight into the man. Garnett was one of three people included in Hudson's will. Through Garnett's wife Constance, Hudson read the classic Russian novelists.

Another greatly influential writer friend was John Galsworthy. In 1904 Hudson sent him a copy of his novel *Green Mansions*, with a covering letter insisting, once again, that 'I've never taken myself seriously as a fictionist'. They had met through Edward Garnett. Conrad called Galsworthy a 'humanitarian humanist'. His series of novels *The Forsyte Saga*, collected in 1922, would become a benign

John Galsworthy

mirror of upper middle-class England that contributed to winning the Nobel Prize in 1932. However, when they first met, Hudson found him 'modern' and a 'little bitter'.

Galsworthy, though, took a step that would change Hudson's economic misery. He sent *Green Mansions* to Alfred Knopf, the cultured publisher, in New York in 1912. Galsworthy's generous and laudatory preface in 1916 would finally earn Hudson a good income from writing, late in his life. William Rothenstein, who drew him, evoked Galsworthy's 'calm, patrician appearance' and claimed he was 'devoted' to Hudson. In 1916, sharing Hudson's animal rights passions, he'd published *A Sheaf*, essays attacking vivisectionists, cruel slaughtering of animals, using horses in mines, even docking horses' tails. The surviving letters confirm how Hudson read his novels and saw plays like *Strife*. Galsworthy visited the sick Hudson in Devon in 1916 where he jotted down in his notebook that

Hudson was 'quite the strangest personality in this age of machines and cheap effects. Like an old sick eagle.'

One hundred and eighty-four letters between Hudson and Margaret Brooke, the Ranee of Sarawak, have survived. She was the most opinionated and upper class of his adoring female friends. They had first met in 1904 and she looked after him at Ascot in 1914 for four months while he was recovering from an illness. She told of how she read *Green Mansions* and was so moved by the poignant last scenes that she had to meet its author. At a dinner party she intuited that a woman guest knew the famously 'aloof' Hudson. She asked to meet him. 'That would be impossible,' she was told, 'he will not meet anybody'. But the dinner guest relented and later telegrammed her when Hudson dropped round for tea.

Once she'd told Hudson that she had lived in Sarawak, he warmed to her. She was twenty-seven years younger than Hudson, had married the far older second white rajah of Sarawak on the island of Borneo, and was reputed to be queenly and attractive. Henry James was a friend. But she was intensely Catholic and denied being 'literary'. She shared birdwatching with Hudson. She found him to be the 'most active man I'd ever met', with piercing eyes, like an eagle's, she wrote in her conventional autobiography. She also found him very modest. She loathed Epstein's *Rima* monument and signed a letter in the *Morning Post* asking for it to be removed.

Another literary friend was the younger nature writer Edward Thomas. They met in 1906 through Edward Garnett at a Tuesday lunch at the Mont Blanc. Hudson was sixty-five years old and Thomas twenty-eight. He could have been a son. Thomas had read him before meeting him, writing a review in the *Daily Chronicle*, 1904, of *Green Mansions*. They rambled together, for example, at Axe Edge on the North Staffordshire Moors in 1913. Forty-six letters from Thomas to Hudson exist, from December 1906 to December 1915.

W. H. Hudson in the New Forest.

Helen, Edward Thomas's devoted and much suffering wife, met Hudson as a widow just once at Whiteleys, under the dome, where Hudson ate every day in his last years, to ask him to write a preface for his late friend. She found him aloof and solitary as he didn't suffer fools gladly. She noticed that his voice was 'very quiet and deep'.

Hudson's great insight into the tall, blue-eyed, lean but self-tortured and depressed Thomas was that this hard-working and tireless writer was 'essentially' a poet, like himself. The difference was that Thomas did start writing poems, under pseudonyms, thanks to his close friendship with Robert Frost in Gloucestershire before the Great War, and had his first collection in proof when he was killed near Arras in 1917 in northern France. Thomas's biographer, Matthew Hollis noted that Thomas admired Hudson more than any other living writer.

Hudson's last piece of writing was the promised foreword to Thomas's *Cloud Castle*, 1922, where he stated that his friend was 'one of the most lovable beings I have ever known'. They were united in both being 'mystics in some degree'. What Hudson meant was that both writers sensed something unnameable, unpredictable, even sacred in nature, that they were nature mystics. In a 1908 letter, contrasting himself to Blunt's Catholicism, Hudson identified himself as a poor creedless mystic. In another letter of 1910 he approved of what William James called him, a pluralistic mystic (I couldn't find that reference). He admitted to enjoying reading James more than any other philosopher. He preferred William to Henry.

Hudson wrote the poem 'The Visionary' for *Selborne Magazine* in 1897, a rhymed account of a 'mental experience re-lived'. He was in the New Forest, in a green leafy world, in silence, when he felt 'unbodied', floating. Hudson wrote that he *knew* the soul was imperishable, defying rationality. In a letter to Roberts he hinted that the source of this mystical knowledge had been during a beautiful blue-sky day on the downs, with viper's bugloss spreading everywhere 'such as I had seen on the pampa'. Then added: 'It was so wonderful a sight that I became the blue of the sky and the bugloss and the air. Why, I didn't seem to walk, I just floated, floated!' Hudson told Roberts in a letter of 1916 that he was a 'religious atheist'. In a letter to Robert Frost of 19 May 1914, Edward Thomas touched on what Hudson meant to him: 'He is, if anything, more than his books.' More than the traces he has left too. And that's my loss and the loss of biography. His writing, fine as it is, only touches the nature of the man himself. No matter how many documents, letters, diaries (and in Hudson's case there's not much), the life itself has gone. There's a ghoulish aspect of writing about the dead, of chasing shadows, of having to imagine that you knew him and can convey something of his spirit.

Sir William Rothenstein's portrait of
W. H. Hudson, National Portrait Gallery.

Some eighty-four letters exist from Hudson to his portraitist Sir William Rothenstein at Harvard's Houghton Library, the earliest from 22 September 1903. Rothenstein (1872–1945), of German-Jewish background, rose to become principal of the Royal College of Art. He was a renowned wit and befriended a diverse bunch of writers, from Tagore, Conrad and Wilde to Verlaine, and wrote the first study of Goya in English. He and his attractive, blonde actress wife Alice lived at 26 Church Row, Hampstead, where sometime between 1910 and 1920, he painted his acute oil portrait of Hudson, as well as countless pencil drawings. The portrait is by far my favourite. It captures Hudson's interior life in his gaze.

Rothenstein and Hudson met at the Chester Square house of Cunningham Grahame's mother. By 1903 they were 'intimate'. Immediately drawn to Hudson, he picked him out as a 'genius'. He noted his poverty, his poor artistic tastes and his loneliness, but 'something about him tore at one's heart, so lovable he was'. He did not define that 'something', though it had to do with 'the animal about him' and his absolute sincerity. Perhaps it was his child-like innocence, as Rothenstein had written in a letter of 29 May 1908: 'You have the rare gift of keeping your innocence.' The older Hudson, enjoying his role of the seer, had warned him about the hazards of sincerity: 'No one who speaks the truth can expect to be believed.'

Nothing like a portrait painter's gaze to fix a sitter's quiddity. Rothenstein never tired of drawing Hudson, even though Hudson didn't like himself as an old man: 'He couldn't bear the idea of growing old and concealed his age.' Well, he hid his age, right from the start. I saw this reluctance in the census forms. Rothenstein noted Hudson's beautifully formed hands, haunting, brown-with-yellow-lights eyes, high, prominent cheekbones and by comparison, narrow jaw. His nose was that of a predatory bird and he had a strange, 'rather crab-like walk'.

So much for his appearance, but Rothenstein and Hudson often met socially, so that the painter sensed Hudson's inner world. Yet, however much he tried, he could never guess Hudson's opinions and could listen to him talk for hours, illuminating the common things of life. That concentration on detail characterized Hudson's world. Rothenstein also singled out a 'strange Spanish pride' in Hudson, meaning being attracted by people whatever their birth. He was as at ease in a shepherd's cottage as in a London salon, echoing Argentinian democratic spontaneity.

As a reader, Rothenstein also promoted Hudson's work. Twice in his memoirs he chose the adjective 'fastidious' about him as a writer. He recorded reactions from Count Harry Kessler who raved about the simplicity, directness and grace of his prose. T. E. Lawrence was another fan and met Hudson through the Rothensteins. Conrad bought one of Rothenstein's sketches of Hudson and noted in a letter: 'You have got the man there in a striking way.'

Two last literary and artistic friends mattered to Hudson in his later years in London. One was Wilfrid Scawen Blunt. In a letter to Violet Hunt, Hudson captured the rebellious, showy poet as 'really a very human and very honourable man', one of the few people he had really loved. This late friendship with the many-sided, anti-imperialist Blunt is surprising. They might have met in October 1905 through S. C. Cokewell, who knew Hudson from the Tuesday lunches at the Mont Blanc in Soho. But it was Cunninghame Graham who cemented the bond. The three recognized each other immediately as genuine eccentrics.

Hudson stayed with Blunt and his wife Lady Anne (Byron's granddaughter) at Newbuildings Place when Blunt, in his flowing white Arabian gown, was an invalid in a chair pulled around by an Arabian horse. Blunt shared Arabian politics and horse passions with Cunninghame Graham and owned an Arabian stud farm at Crabbet Park. In what seems like fancy-dress today, Blunt acted out the revolutionary orientalist, supporting freedom in Egypt and Ireland. He was imprisoned in 1888. He was also a poet, famous for his affair with Jane Morris, and was constantly in debt. So Hudson saw beneath the mask, but didn't share Blunt's public opposition to the British Empire. In a letter, he wrote that he endured 'endless discussions of Arabian politics'.

However, they were united by being the same age and both being outsiders. Blunt was a Catholic who refused to go to university. Instead he became a diplomat in Madrid for a year, where he trained as a matador. He was posted to Buenos Aires in 1868, again for a year, while Hudson was still there. He told a friend that he 'talked Spanish fluently' and 'adored' all Spanish things. Blunt bumped into Richard Burton down from Brazil, a heavy drinker armed with a gaucho knife, who had stirred up his desire to go to Arabia. In his diaries Blunt found Buenos Aires suffering an 'atmosphere of

Drawing of W. H. Davies by
Dame Laura Knight.

violence', with men shot in the street nightly. Hudson had met another writer, like Cunninghame Graham, who spoke Spanish and knew Argentina. Ford grouped the three as 'Cunny G, Hudson and Wilfred [*sic*] Blunt'. They were often together, three childless writers who admired each other greatly.

The other person was W. H. Davies, who was brought by Edward Thomas to the reserved table on the second floor of the Mont Blanc restaurant on a Tuesday. He was shy, with a reputation of having lived as a tramp in the United States and England, but relieved that Hudson was always the first to arrive at the table. He was surprised at Hudson's 'discernment' in recommending a good poet to him, but hadn't read any of Hudson's books. At this first meeting he sensed Hudson was embittered by old age, and not being recognized as a writer.

Davies's love of birds probably exceeded Hudson's, he thought, but he admired Hudson's 'extraordinary knowledge' of bird life and habits. He was equally impressed by Hudson's literary side: 'How he managed to read so many books and magazines I cannot say. Nothing seemed to escape him and everything seemed to be read with a critical eye.' But he managed to make him laugh. When Belloc asked Hudson for a piece for a new magazine, Hudson said: 'What do you pay?' to Davies's surprise. It wasn't until Conrad said that 'Hudson is a giant' that Davies got hold of *Green Mansions* (he couldn't afford books) 'and found it one of the most fascinating books I have ever read', while he couldn't finish any of Conrad's 'melodramatic' novels. Then Davies, after one lunch, set off for a fortnight's walk, with a just a few bits and bobs in a pocket and no books.

Chapter 13

Rural England and the Industrial Mills

Hudson was born in such a primitive area on the pampas in 1841 that it remained untouched by progress. In 1874 he travelled to England and never returned, but he couldn't abandon his vivid experiences at his birth farm. Over the years until his death, he began a slow conquest of rural England, made even slower by the distance he had to catch up, and by being a foreigner.

As he grasped rural England, it was slipping out of view. By the 1851 census, England had ceased to be a rural country as more than 50 per cent of the population lived in towns. However, Hudson never really healed the wound between his savage past and his English present. His writing attests to this clash between the apparent simplicity of savage life and the sophistication of modern urban values. As a writer he appealed to two sets of readers: those nostalgic for rural realities and those excited by modern sensibility. I sought to explore his extraordinary process of uneven assimilation between 1874 and 1922 (his death) through confusions in his writing. Hudson wrote a series of books with catchy rural titles like *Hampshire Days*, *A Shepherd's Life*, *Afoot in England* and *Nature in Downland* that were read, in Tippett's phrase, as 'the quintessence of Englishness'. But that really meant, 'rural Englishness'.

Ralph Waldo Emerson, in his *English Traits*, 1865, found the uncultured English a brutal nation, adoring boxing, bear-baiting, cock-fighting, flogging and watching executions. England had shrunk to a garden, often walled, with countless grand houses preserved by primogeniture and endless wealth, an 'all-preserving island'. The feudal system still existed. Class in England was allied to an absolute homage to wealth. An insult was to call someone a 'beggar'. It was a disgrace to have been born poor. Hudson suffered his pampas origins in silence.

Emerson pointed his finger at the English love of home, domesticity and privacy – what he called their 'incuriosity'. Taciturnity ruled the English, kept their lips sealed. Understatement doesn't allow for intimacy. The moment you open your mouth, some Englishman hates you, wrote George Bernard Shaw.

Hudson struggled to grasp this 'England'. However, Emerson also picked out rural lore: love for country lanes, farmyards and markets. He noted how the English avoided humbug and pretension, and resisted change. Here was an 'England' to which Hudson could relate. An unchangeable place, except in London. London, in Emerson's words, was the new Rome of the mid-nineteenth-century, where pineapples and oranges were cheaper than in the Mediterranean. Industry was king, where the English perfected things ill-made elsewhere, like weaving ponchos for the Mexicans. The steam driving machine 'is almost an Englishman'. But London also had the darkness of its sky, its fog and fine soot, so vivid in Dickens's opening to *Bleak House*. England was a ship anchored off Europe and the English 'cannot readily see beyond England'.

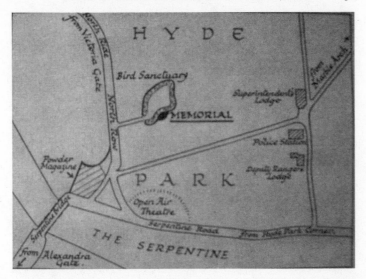

Invitation map to the opening of Hudson's Hyde Park Memorial, 1925.

After arriving in England, Hudson travelled once to County Wicklow in August 1896, where his Irish paternal grandmother was born a Maloney (Hudson got her name wrong as Moran) but never crossed the Channel to mainland Europe. He almost became a little Englander.

Hudson had to quietly strip himself of his Argentine background. He praised Edith Wharton for 'de-Americanizing herself'. Henry James also stayed in England and Hudson admired his 'intense Englishness'. Just before his death, James applied for English nationality. He was a passionate pilgrim, a grateful alien. James was almost Hudson's exact contemporary (1843–1916) and they met, though Hudson never asked him why he chose the surname Hudson for one of his novels.

Like Hudson, he knew London through literature before arriving. He found the city a 'strangely mingled monster', where, for an American, history survived to 'vibrate through my mind'. Though aspects of this murky, modern Babylon were hellish – Hudson would agree – with 'miles upon miles of the dreariest, stodgiest commonness. Thousands of acres covered by low black houses of the cheapest construction, without ornament, without grace, without character or even identity', yet James remained a London-lover. Londoners did not idle and clocks ruled social life. There were crowds everywhere, so that James began to admire the high walls shutting others out. He joined those inside, the cultured and the aristocratic, in order to find intimacy. His initial 'horrible' journey from Euston Station to the Morley Hotel in Trafalgar Square was quickly forgotten. The creak of a waiter's shoes, for example, became the sound of tradition, much like the taste of muffin.

The best expression of this urban love for rural England comes from the three-times Conservative Prime Minister Stanley Baldwin, who opened the Hudson Memorial in Hyde Park. In a 1924 speech called 'England', he begged the English not to 'ape any foreign country'. England has a 'knack of producing geniuses'. In fact, 'the Englishman is made for a time of crisis' and is 'ruthless in action'. Baldwin is proud that this Englishman is 'impervious' to intellectual impressions. He enumerates English sympathy for the underdog, the ability to laugh and 'diversified individuality'. He turns to rural England – 'the sound of the scythe against the whetstone' – and the English 'love of home'. Hudson had to be very cautious. In a letter to Cunninghame Graham in 1894 he knew that the English reader had far less interest in gauchos (or Argentines) than in any people or races within the British Empire.

Initiation into English society was never a subject Hudson broached directly. But it's the secret theme of his second, anonymous novel *A Crystal Age*, 1887. His protagonist Smith has to shed his English superiority as he adapts to a new society that ignored England, then the greatest empire on earth. Hudson understood this imperial superiority when in *Land's End*, 1908, he wrote about the 'fact that the Englishman is endowed with a very great idea of himself, of the absolute rightness of his philosophy of life . . . He is, so to speak, his own standard and measures everybody from China to Peru by it.'

Even the typical surname 'Smith' meant nothing. So Smith tries out famous English surnames on the old man who rescued him: 'Well, I suppose you have heard of some of my great countrymen: Beaconsfield, Gladstone, Darwin, Burne-Jones, Ruskin, Queen Victoria, Tennyson . . .'(and the list goes on). But not one name rang a bell. Hudson had included an epigraph from Darwin's *Origin of Species*, which he removed in later editions, about a 'prophetic glance into futurity' where natural selection will ensure the human being's 'progress towards perfection'. As Smith tries to fit in to this future society he finds his Londoner's garments 'uncouth' and his upperclass English a 'thick speech'. He is shamed and has become a blunderer. Smith devised a tactic that was Hudson's in the late 1870s:

> Of course I was surrounded with mysteries, being in the house but not of it, to the manner born; and I had already arrived at the conclusion that these mysteries could only be known to met through reading, once that accomplishment was mine. For it seemed rather a dangerous thing to ask questions, since the most innocent interrogatory might be taken as an offence . . . To be reticent, observant, and studious was a safe plan . . .

It couldn't be clearer. Hudson, not born in England, adapted by reading in the British Library, observing and not opening his mouth. At one moment, Smith curses England, just as Hudson had done: 'Oh, that island, that island! Why can't I forget its miserable customs . . .' But finally, Smith celebrates that joyful day 'when I was to cease, in outwards appearances at all events, to be an alien'. Hudson too would cease to be an alien, at least on the outside, but unlike Smith, did not commit suicide. Instead, he stole away into a grove of stately old trees and sat on a large twisted root:

> To this sequestered spot I had come to indulge my resentful feelings: for here I could speak out my bitterness aloud, if I felt so minded, where there were no witnesses to hear me. I had restrained those unmanly tears . . . and kept back by dark thoughts on the way; now I was sitting quietly by myself, safe from observation . . .

Roaming alone, unwatched, in nature was Hudson's medicine. His friend Roberts noted that Hudson was reluctant 'to admit that he was not English by birth' so dire was the experience of being an alien.

Hudson's struggle with English society, its classes and psychological traits has moved on from Emerson, but not that much.

When my family arrived in London from tropical Mauritius in 1950, we all suffered the climate, the rationing and the secret codes of behaviour. After my mother's death, I came across her bible on how to behave, Nancy Mitford's *Noblesse Oblige*, bought in 1956 and mercilessly applied to her four children. She was constantly correcting us, so that I still pronounce forehead fully, knowing it would grate on her ears. And she was Norwegian. This little booklet doesn't give me any special insight into Hudson, but does remind me of how Englishness as subtle codes penetrates everything. And how difficult it was for my colonial artist father to feel at home in London. He had few friends. His social life was directed by my mother. He buried himself in his action painting, in his intellectual

pursuits and in his family. He was literally a family man. He'd published a book of poems in Mauritius in 1949. When he arrived in
London he was shocked at how dated these heartfelt poems were.
He was still Tennysonian, while English letters were into the
'Movement', in a post-Auden phase. He stopped writing, almost in
shame. He turned instead to his abstract, speculative, scientific ideas.
He wanted to combine metaphysics with physics and talked of
'string theory' or 'crystals', but at a human level, he was always
trying to convince me. I resisted. And we talked and talked. Late in
his life, my mother would beg me to 'talk' to him. He was an emotional man, but I feel now all this pent-up, private self was a
consequence of being a colonial exile in England. He sought to be
rooted in his mind, not a place.

And London is where Hudson found himself. In 1892 Hudson
published a three-volumed novel *Fan, the Story of a Young Girl's Life*,
under a pseudonym, Henry Harford. Henry was his second name
and Harford is both a Devon and Maine village. Biographer Alicia
Jurado saw a marble plaque in Exeter Cathedral saying 'Luisa, Wife
of Henry Harford', and Fan, his protagonist, did visit Exeter
Cathedral in the novel. But why did Hudson use a pseudonym?
Hudson was struggling to make money through writing, but didn't
want anybody to confuse the fiction writer with the field naturalist
who, at the same date of 1892, published *A Naturalist in La Plata*.
Only 350 copies of *Fan* were printed. Alicia Jurado saw one copy
where Hudson had jokingly dedicated his novel to himself, from
Henry Harford to W. H. Hudson. This realist novel, with accurately
observed dialogues and constructed on coincidental meetings,
follows Fan out of London poverty, a beggar-girl-to-riches tale. It's
a Dickensian pastiche, a potboiler. He told his publisher to never
reissue it, but then repented, and J. M. Dent republished it
posthumously in 1923.

As biographer Ruth Tomalin noted, the background is Hudson's London slums off the Edgware Road, and fifteen-year-old Fan is a faint double of Hudson. She emerges from extreme poverty and hunger because she is an orphan rescued by an eccentric, piano-playing woman (Miss Starbrow is based on his musician wife Emily) – she seems refined, despite her social origins. Fan longs for the solace of nature and finds it in Kensington Gardens, like Hudson himself. As an orphan without any family or social support, she slowly makes her way into respectability. Hudson walked the same dingy streets as she did, was a foreigner/orphan, taken up by notable, well-born people. You even hear the church bells of St Matthew's ring, the church where Hudson and Emily married.

When Fan is sent out to be tutored in Eyethorne, Wiltshire, she discovered the therapy of wildflowers. Her tutor is a clever, godless young woman, rebelling against her church-going mother. Hudson, through these female characters, relives his own inner struggle over Darwin and religion with his mother. Fan, so like Hudson, resorts to the 'mystic silence' of nature to escape London stress and poverty. In a wood she feels a 'great love that we can hardly bear it', something deeper than human love that remains beyond rational analysis.

Hudson's extensive London rambles, his poverty and his being a foreigner can be read back into this novel through Fan. Below the plot, there are authorial passages about nature, love, friendship or crushes between women, religion and a curiosity about Londoners. Most telling is how Fan reacts to 'tyrant Memory', for it was in the 1880s in London that memories of his forgotten Argentine past kept Hudson alive. Here is his version of this consolation that would lead to *Far Away and Long Ago*:

For now a hundred sweet memories rushed into her heart – her walks in the Gardens, all the little incidents, the early blissful days when she lived with Mary, and so vividly was the past seen and realised, yet so immeasurably far did it seem to her and so irrecoverably lost, that the sweetness was over-mastered by the pain, and the pain was like anguish.

Hudson mocked himself as aspiring novelist through the struggling writer Constance, married to an intellectual idler, and also living in poverty in Mile End. A publisher had accepted her moderately good novel, but it was in only one volume, when the public preferred three volumes. As an unknown author, she had to accept a pittance. She was reluctant to compete with other novelists. She wanted to withdraw the novel, improve it, imagining how critics would tear into it. Her curate friend thought the novel was too well-written, almost amateurish: writing a novel was not literature and rewriting was not acceptable. To be an unknown novelist was to live on bread and cheese. So Hudson, within his novel and through Constance, portrayed his own creative turmoils.

Hudson was born poor on his pampas farm. He told Professor Spencer Baird in his first letter of 1866 he was not 'a person of means'. We know exactly that Hudson earned ninety cents for each bird skin, but also that it wasn't sufficient, that he had no leisure to collect and his payments scarcely met his expenses. He almost gave up bird collecting. Then in England there was a black week with Emily in Ravenscourt Park, surviving on cocoa. He began earning as a writer or journalist, as he put it in the censuses, but being a professional writer was constant penury. Early on, he told his mentor Dr Sclater in an unpublished letter at the Zoological Society that he had been rejected by a publisher for a planned book on Buenos Aires's zoology and advised 'not to attempt such a thing'. Instead, he should publish articles in the *Gentleman's Magazine*. The world of Fan was his own. Perhaps also, he wasn't scientific enough for a book. For example, *A Crystal Age* didn't earn him a halfpenny. He was also threatened with being sued for repayment of his £25 advance by publishers Chapman Hall in 1897 for not completing a book on Argentine birds. In *Afoot in England* he moaned: 'But I fancy the nearest crossing-sweeper did better and could afford to give himself a more generous dinner.'

Such was his breadline survival that his powerful friend Lord Grey, encouraged by Cunninghame Graham, managed to get Hudson on the Civil List Pension worth £150 a year. This fund, set up by Queen Victoria in 1837 for discoveries in science and the arts, was in Hudson's case recognition for the 'originality of his writings on Natural History'. But to get it, Hudson had to take out British nationality, which he did in 1900. Just before he died, in 1921, he resigned his Civil List Pension. In 1916 £200 was collected amongst friends by Garnett and placed in his account. Rothenstein confirmed that Hudson never discovered the secret. Hudson was only financially comfortable when *Green Mansions* started selling well in the 1916 American edition (there were over seventy editions by 1977), promoted by Galsworthy's foreword comparing him to Tolstoy. Knopf, his publisher, recalled years later that he fell in love with the book as a student and that it 'launched our firm'. That it had done so well, wrote Knopf in 1917, was a 'source of greater satisfaction to me personally than anything that has happened in connection with my business'. In a letter of 1903 to Roberts, Hudson stated his case:

What revolts me is the thought that when I had not a penny and almost went down on my knees to editors, publishers and literary agents I couldn't get a civil word . . . And now that I don't want the beastly money and care nothing for fame and am sick and tired of the whole thing they actually come to beg for a book or article from me.

W. H. Hudson by Marie Leon.

Richard Curle claimed that there 'never was a man who cared less for money'. This dismissal of the importance of money was the downfall of his father, a family trait.

After twenty-six years in England, Hudson decided to naturalize himself English. He was being recommended for the Civil Pensions List, had to be British and needed the income. Ornithologist Philip L. Sclater, entomologist Charles Owen Waterhouse and Morley Roberts proposed him. In the modern National Archives building in Kew, in a heavy bound volume of naturalization documents, in Hudson's legible handwriting in fountain pen, he states that he was 'a subject of the Argentine Republic and was born at Quilmes, Buenos Ayres', and that he was the son of Daniel and Caroline Hudson, 'both citizens of the United States of America' (but not that they had died). He was a fifty-nine-year-old 'author and journalist', married, without children. On 7 July his Oath of Allegiance was confirmed. England was not an immigrant country, like the United States and Argentina. So immigrants like Hudson stood out.

Wells from the air.

He commented in a letter: 'It is no easy matter to become a citizen of this small country'.

The only reaction to this oath was in his article 'The Feather Tradition', in the *Humane Review* of 1901 where he proudly called himself an Englishman. In a letter to John Galsworthy, Hudson identified with Henry James, whom he met once or twice: 'I am grateful to him for his books, but even more for his having made himself an Englishman at the end of his life . . .' Henry James was another American who'd renounced his birth country.

So was T. S. Eliot, who took the same oath in November 1927. To become an English citizen and to be received into the English Church would remove the stain of being a 'squatter'. In a letter to his brother, Eliot revealed he had needed to pull a few strings at the Home Office, but hoped his brother wouldn't be shocked and 'don't tell mother'. So naturalization could also be a betrayal. Peter Ackroyd saw Eliot as a 'resident alien', both at home and not at home in England. Did naturalization fulfil Hudson's dreams, as it did for Eliot and James, and did he also remain an alien?

In 1920 in his revised *Birds of La Plata* Hudson wrote negatively about his choice of becoming English: 'When I think of that land so rich in bird life . . . I probably made choice of the wrong road of the two then open to me.' In 1949 there was an attempt to repatriate Hudson's remains back to Buenos Aires (as Gardel's and Rosas's were), but it came to nothing. It's impossible to decide what Hudson really felt, except that naturalization brought him money and status. I also had to naturalize myself in 1968 when Mauritius gained its independence, but it meant little. I thought of myself as English and my first wife was English. I hadn't yet discovered that my English grandmother wasn't my grandmother and that I hadn't a drop of English blood. To me naturalization was just a bit of tedious paperwork, and only my elder sister remained Mauritian. My intense childhood memories seemed remote and my life in London was cosmopolitan. Only exploring Argentina and Hudson made my life seem more confused than I thought.

Becoming an Englishman overnight through naturalization in 1900 is a landmark date in Hudson's private calendar. He could write for English readers as if he was one of them.

His Englishness was subtle but I sought it out in the special qualities of the county towns he wrote about, even though David R. Dewar claimed that Hudson's England can only be found in his books. I take it as a given that reading him is an act of nostalgia for a lost England. Yet there are corners where change has been imperceptible. One is Flushing. He wrote to his American publisher Knopf about a corner of England that hadn't changed since the thirteenth-century, a little huddled village against a cliff, with everything 'exceedingly primitive' (something for which Hudson looked).

Another is the cathedral and its green at Wells in Somerset. During his first visit in 1895, Hudson wrote that he preferred Wells to any other county town. In April or May, standing on a neighbouring hill, he would 'sigh with pleasure'. He particularly liked the way nature surrounded the town, breathing on it, touching it. He fancied 'you will not hear a green woodpecker in any other English city'. He took a room on the green and observed the magnificent cathedral.

In 1905 he wrote that this 'loveliest' cathedral in the country had a west front, with its truncated towers, so richly decorated that it gave refuge to over 200 jackdaws. Here's Hudson at his most peculiar – it's because of the birds that he adored the place. He would spend hours watching them caw and squabble in their nests behind the statues. He listened so intensely that he boasted, 'I knew just what they were saying.' For Hudson found that jackdaws were one of the most intelligent of 'our' birds. That ability to concentrate for hours on a bird developed from his boy's desire to see birds so closely that the intensity was 'almost a pain'.

Contrast him with Henry James, who was pleasantly shocked to find a great cathedral without a city, looming over the countryside, with no factory chimneys in sight. For him the west front is the cathedral's great 'boast', with over 300 statues, densely embroidered by the chisel. But he doesn't remark on the equally stunning great east window, smashed by Thomas Cromwell, with the glass pieces collected and put back higgledy-piggledy, though I noted an arm and a halo intact when there.

My father took us as boys to this east window and pointed it out as a great abstract expressionist painting (he was Frank Avray Wilson, a pioneer *tachiste* painter in England). Hudson had appreciated his favourite cathedral window as 'without design', an apt term for the way he wrote and rambled, without conscious plot. Hudson stayed in Wells three further recorded times. A biographer,

Dennis Shrubsall, also lived there and confirmed in 1978 that the 'Wells of today is little more than the market town it was in Hudson's time'. It's not impossible to repeat Hudson's experiences on the cathedral green.

Hudson turned landscape and life into dreamy stillness. He lifted a place like Cookham Dean, near Maidenhead, out of history, but held onto its vivid, secret life. In 1892 Hudson lodged in Midway Cottage on Pope's Lane in a hollow surrounded by a common, orchards, woods and the Thames (he made three further recorded visits). He wrote about his two months from May to July without naming the village. He talked with locals in the pub, caught urchins stealing eggs, spied on bird catchers and grumbled about gamekeepers. His stay was probably his first outside London, due to money problems, but after a hard winter, he found this little paradise. He would get up at about four at dawn and listen to cuckoos by his cottage window. Then he would saunter off with his binoculars, spotting fifty-nine species. One lunch was a cake, a banana and whisky from his pocket flask. Hearing nightingales singing in daylight became his medicine; pure, refreshing melody.

He escaped not only the city in actuality, but also when he sat and wrote, so that his readers can also escape by *reading* him. His soothing prose captures what is normal, but Hudson is seeing typical village birds for the first time, like an explorer in the shires rather than in Surinam, or as a *TLS* reviewer put it in 1912, 'Mr Hudson is still discovering England with half-foreign eyes'. As Hudson noted, local villagers were hopeless observers, numbed by familiarity, unlike his secret foreigner's view. In the long section on Cookham Dean in *Birds in Town and Village* (the second edition of his first English book of 1893), Hudson found his ideal spot under some elms and yearned for the freedom of being a boy again, to climb, swim and revel in the sun.

Then, as he is actually writing about this sequestered spot in London, Hudson would have asked his favourite poet, Mélendez Valdés to be quiet and listen to a tree-pipit (art is always secondary to life). The effect of its canary-like trills and thin piping notes infect the mind with a wish 'to be perfectly still and drink of the same sweetness again and again in larger measure'. To imitate Hudson, find your little patch and listen, as he led me to do. Or even better, read him as another version of Wordsworth's emotion recollected in tranquillity.

It was at Cookham Dean, or during a digression in his writing about it, that he imagined a fairy-tale similar to the Rima one in *Green Mansions*, and thought about cruelty and then the 'awful fact of death'. Here an Argentine memory burst into the English countryside while he wrote – or while he meandered in his thoughts – concerning a purple martin he watched sporting in the sky that suddenly crashed to the ground at his feet. He dissected it to confirm that it had died from the 'natural failure of the life-energy'. But with human beings such a way of dying without a struggle, without a pang, that wild creatures enjoy, is rare. Hudson, in his rambling essay, infused birdwatching with human watching, with a plea for meditation about how to die.

What changed Hudson's attitudes to rural England was the modern invention of the two-wheeled bike. So he wasn't averse to technology. Morley Roberts introduced Hudson to the bicycle. Too poor to own a horse or a car, the silent bike was ideal. A bike is close to a horse, the saddle, the brakes, legs peddling on each side. In a letter of 4 February 1910, Hudson, aged sixty-nine, wrote: 'The bicycle suits me near enough and after a few miles runs I feel that I have breathed deeper and got more red blood than if I had walked.' The silent bike allowed Hudson stealth. Of course, it broke down. Once a thorn punctured a wheel. Another time in 1908, he biked to

Stonehenge at midnight to catch dawn and was surrounded by five or six hundred people who had arrived on bikes.

In *Afoot in England*, he wrote:

> Two or three season ago I was so unfortunate as to run over a large and beautiful bright grass snake near Aldermaston, once a snake sanctuary. He writhed and wriggled in the road as if I had broken his back, but on picking him up I was pleased to find that my wind-inflated rubber tyre had not, like the brazen chariot wheel, crushed his delicate vertebrae: he quickly recovered and when released glided swiftly and easily into cover.

Hudson's chain-driven Rover cost him £10 in 1890 and was ideal for his birding. In a whimsical, late piece on annoying wasps, Hudson launched into how bikes helped liberate women once the male shock of seeing them astride wheels was overcome. My experience of bike-riding in London, along side streets and avoiding bus lanes, would begin with that wonderful sense of freedom and alertness it gives.

In his mature years, Hudson as a writer seemed to glow in the pride of having become a naturalized Englishman, or perhaps in the nature of the country he had by then long inhabited. This is not in chronological order, but captures his best nature writing in four books written in his sixties. He almost created his own genre, based on digressions, so that he could write of anything within the format of an essay.

Hudson once recalled a local shepherd on the pampas, who was really an illiterate gaucho. But this experienced man taught Hudson a crucial lesson. After a local boy had humiliated him, Hudson wanted his parents to take revenge. But the old shepherd advised him to take his time and wreak his own revenge, which he did. The message was Machiavellian; do not rely on others, but use cunning to get your own revenge. It was a life lesson, poignant for an exile without a home network. In 1910, in a letter to Cunninghame Graham, Hudson found the old men, often gauchos, evoked in the story 'El Ombú' came from a 'world more interesting than our best shepherds'. With this insight about wise, illiterate gaucho shepherds in mind, Hudson set about working hard on a life of an English shepherd.

He carried out his task so well that Ezra Pound, in 1920, concluded *A Shepherd's Life* was 'art of a very high order'. He praised Hudson's work as 'so true to the English countryside' (though what Pound knew about it as an American could be questioned). As rural traditions become forgotten, Hudson's modest shepherds are made to speak for the loss of a 'life of simple, unchanging actions and of habits that are like instincts, of hard labour in the sun and wind and rain . . . and but few comforts and no luxuries'.

Here we have in a nutshell Hudson's own life ethic, learnt on the pampas and forgotten in comfortable, urban, Edwardian England: only the rural workers still belong to the earth. Hudson did too, but at a cost, for he had chosen the Downs, so like the pampas, for a 'sense of being at home wherever grass grows'. Throughout his study of shepherds was the backdrop of emptiness and silence in a wilder nature. Hudson longed to 'hear and see and feel the tempests of rain'. Writing about English shepherds was a secret therapy.

When his book finally appeared in 1910, despite an essay on 'A Shepherd on the Downs' in *Longman's Magazine* in 1902, Hudson hid his main character James Lawes under the pseudonym Caleb Bawcombe, and the villages of Martin under Winterbourne Bishop and Doveton under Upton Lovell (much as Hardy had done in his Wessex). Morley Roberts tried to force Hudson to let the cat out of the bag, but Hudson protected his shepherds. He had recorded their

William Lawes, shepherd, and his wife.

lives in now lost notebooks. Their anecdotes became the social history of the downtrodden. He was an old-fashioned anthropologist, a listener. Just as he learned from gauchos round the fire (it was not his own experience that mattered), so this anti-egoistic writer just listened. He also culled newspapers and reports from the Assizes. He reached back into the memories of nonagenarians to 1830 and Luddite mobs smashing the new threshing machines.

Hudson met Lawes in 1901 when he chanced to lodge in 'The Pines', a thatched, timber-framed cottage, now a bungalow. He had biked nine hours from Salisbury and had written to his RSPB friend Mrs Hubbard about 'one of the loveliest out-of-the way spots I have seen. This is all a county of vast Downs, where you can scarcely see a house and can find yourself in a wilderness'. It was the opposite of picturesque, and the humble houses were poor, without flowers. Hudson struggled to overcome Lawes's shyness and reluctance to speak. He was lucky in that he too knew poverty, the tough farm-hand life, and was not a landowner or squire. In a 1908 letter to

Garnett he wrote: 'I detest . . . the upper class'. In fact, shepherd Caleb secretly mirrored Hudson.

The book opens with an elegy to Salisbury, its cathedral and its market day from a foreigner's perspective. He listened, quoted and noticed, amused by people. Then he moved on to Martin, the least attractive village on the Downs, but the most appealing to Hudson, exactly due to its lack of charm. It was cold, waterless, treeless, unblessed: Hudson piled up negatives. Nobody wanted to visit, and few strangers turned up. Here he fell in with his shepherd. This was real England, not the land rich in ancient monuments and grand parks and historical buildings. Hudson's justification for his choosing Martin: 'That emptiness seemed good for both mind and body: I could spend long hours idly sauntering or sitting or lying on turf, thinking of nothing, or only one thing – that it was a relief to have no thought about anything.'

Hudson is almost Buddhist in his view of the mind and the verb 'saunter' derives from 'saint' and visionary musing. Of course, a shepherd couldn't do this, but he had hit on a secret reason: this empty place was 'more like the home of my early years than any other place known to me in the country'. Here he was no longer an intruder, but belonged with the meek and powerless as part of an age-old way of life on the verge of extinction.

Caleb had retired. He was lame, seventy-two and the son of a shepherd. Hudson captures his exact peasant-talk ('We be strangers to he'). Caleb began looking after flocks aged six. He was tall, recalled all his sheep dogs, as well as poaching and the terrible fate of sheep-stealers (sometimes hanged, usually sent to Tasmania). Caleb was taught to read, but the sole book he owned was the Bible, which he carried around with him. At this point, Hudson realized that the ancient stories he was noting down were just like ones from his childhood pampas. Both Argentine and English peasants lived

like patriarchs, 'old, sober, slow-minded herders of the plains'. In such a densely populated and civilized country like England, the 'ancient spirit had survived'.

Hudson, faceless, called himself an 'untrammelled' person, in exile or captivity in the city. He outlined his researcher's technique. When he sat by the fireside and smoked with Caleb, he couldn't ask direct questions. Caleb would ramble and suddenly, by chance, unleash a memory. Hudson had learnt his 'mental habit' of patience as a field naturalist. Always watchful, never in a hurry, so Caleb unwound his life digressively over the nine years of their meetings. When Hudson mentioned hedgehogs, Caleb remembered people who hunted and roasted them on the spot, begging salt from a nearby house. Caleb admitted that he dreamed always about sheep. The book ends with Caleb's posthumous wish: 'Give me my Wiltsheer Downs again and let me be a shepherd there all my life', not very different to Hudson's own wish about the pampas.

I revert to my opening exploration and William Lawes's tomb (he was Caleb's father) in Martin with its 1952 inscription: 'A shepherd of the Wiltshire Downs. William Lawes was the Isaac Bawcombe of W. H. Hudson'. It could have been Hudson's own epitaph.

The South or Sussex Downs mirrored what Hudson found most attractive in England: 'Treeless they were, and if not exactly repelling . . . uninviting in their naked barren aspect'. Yet, despite this uninviting background, his contemporaries quickly acclaimed his next book, *Nature in Downland*, as his masterpiece. By 1906 it had been reprinted three times. Ford Madox Ford, author of a quickly written trilogy on the English countryside, found it the 'most beautiful book that was ever written about an English county'. He said he knew the opening paragraph by heart. It encapsulates Hudson's peculiar version of England:

On one of the hottest days in August of this exceptionally hot year of 1899, I spent a good many hours on the top of Kingston Hill, near Lewes. There are clear mornings, especially in the autumn months, when magnificent views of the surrounding country can be had from the top of that very long hill. Usually on hot summer days the prospect, with the sea of downland and the grey glinting ocean beyond on one side, the immense expanse of the wooded Sussex weald on the other, is covered with a blue obscuring haze, and this hot, windy August day was no exception. The wind, moreover, was so violent that all winged life, whether of bird or insect, had been driven into hiding and such scanty shade as existed; it was a labour even to walk against the wind. In spite of these drawbacks, and of the everywhere brown parched aspect of nature, I had there some hours of rare pleasure, felt all the more because it had not been looked for.

Hudson, alone and patient, relished the wild wind; there's his contrariness and a strange tranquillity. He wrote the first chapter in Jefferies's house in Goring, some parts in Brighton Library and the rest in his tower in Westbourne Grove, so it's also a memory, but so intense he and his reader re-experience this hilltop vision. Nothing in his style distorts what he's looking at, yet half the pleasure is the rhythm, his calm voice. The rest of this book follows from this opening as a wind from the past, a raw pampas blast.

What motivates Hudson both as rambler and writer is 'aimlessly wandering'. All his late books are experiences of randomness. Digression is his mainstay; the phrase 'to get back' structures his mental and rambler's universe. And digression is a metaphor for spontaneity and freedom.

By 1900 Hudson is sure that he's not a 'frigid' ornithologist and he's aware that he is creating a new kind of writing. This is a 'small unimportant book, not entertaining enough for those who read for pleasure only, nor sufficiently scientific and crammed with facts for readers who thirst after knowledge'. He has not written a guidebook, not even a travel book. In fact, Shrubsall pointed out that Hudson 'never wrote a complete account of a journey'. He kept no journal. These are casual meditations on 'spiritual geography'.

Hudson is aware of his literary persona: 'On this subject I therefore speak as a fool, or, at all events, an ignorant person.' The holy fool implies an outsider's tradition, with William Blake at its core. It's a foreigner's angle on quintessential England. Behind all his writing we sense his commitment in five short, italicized words, 'what we see we feel'. So he bundles together comments on the Downs' wild or fairy flowers like bindweed, its insects, especially flies, foxes, and much on adders, as well as on Hurdis, Wells Cathedral, Gerarde, De Quincey, birdsong and a girl biting into a green apple. For, despite all our scientific knowledge, 'we do not and never can know what an insect knows or feel what it feels'.

The third volume of collected sketches in this Downland trilogy is *Hampshire Days*, dedicated to his powerful RSPB friends Sir Edward and Lady Grey 'with Hampshire in their hearts'. In a letter of 1904, Hudson told Dr Sclater that he picked on Hampshire because it was 'so rich in wild life'. Edward Thomas thought Hudson was 'most at home in Hampshire', but didn't go into it much further. And many later nature-starved town dwellers would agree with poet P. J. Kavanagh that *Hampshire Days* is his best book. Although Hampshire lends unity to the book's rambling method, I feel that Hudson has found his form – that is, loosely connected, already published sketches. He called this a string of 'fragments'. From the early 1900s there is no further development. He continued to write one long, ongoing book until his death in 1922.

Hudson's development as a writer goes against the grain in that he blossomed as a writer when over sixty years old, akin to Cervantes who published his first part of the *Quixote* when fifty-eight and the second when sixty-eight. I keep a list of late-developing writers as a talisman. Violet Hunt saw Hudson as many did: a 'graceful, perfect old man of unknown age' as if he and his Downs landscapes were ageless.

Just as Hudson rambled so does his writing. He started in 1900 in the New Forest and passed through Beaulieu, then White's Selborne, far preferring the hilltop common, 'wildest in Britain', to the actual village. After Wolmer, he stayed at the Greys' cottage in the Itchen valley. So, obviously it's not a guidebook to all Hampshire. As usual, Hudson is alert to the 'little incidents' as he patiently noted down every variety of life from flies and stag beetles to woodpeckers and a long section on how cuckoos ejected fledglings from a nest, a task that A. R. Wallace had asked him to observe. There's a chapter on Hampshire types.

But everything is reflected in Hudson's mind as he wrote. 'We live,' he said, 'in thoughts and feelings, not in days and years.' Inside his mind there's no history, no chronology; it's the relief of an illusion of timelessness, or to use his own favourite word, unchange-ableness. You read Hudson to escape urban restlessness. He sought little crannies of wilderness and found them in what he called 'the rude incult heath, the beautiful desolation; to have harsh furze and ling and bramble and bracken'. That was his real 'territory' where he refreshed himself.

Ruth Tomalin called this his 'forest' book as he was also writing about the jungle in *Green Mansions*. He gets so close to life that adders 'hadn't told me a hundredth part of their unwritten history'; as if he was being dictated to by adders. No wonder Hudson found nature superior to any human art, for it spoke to him in his mind. Gazing at an oak in August would make any artist 'sick of his poor little ineffectual art'. He sounds like D. H. Lawrence. Then, sud-denly, an outburst from his Argentine past concerning spiders, that rubs in his foreignness, despite having become English in 1900:

It made me miserable to think that I had left, thousands of miles away, a world of spiders exceeding in size, variety of shape and beauty and richness of colouring those I found here – surpassing them, too, in the marvellousness of their habits and that ferocity of disposition which is without parallel in nature. I wished I could drop this burden of years so as to go back to them, to spend half a lifetime in finding out some of their fascinating secrets. Finally, I envied those who in future years will grow up in that green continent, with his passion in their hearts, and have the happiness which I have missed.

In 1920 Hudson published a simple, gripping Saxon fantasy called *Dead Man's Plack* where queen Elfrida, exhilarated on horseback, echoed Hudson's Down-love. We can read him back into his char-acter as she escapes the mundane world and saw the prospect of the

Downs 'sharp and clear as it had never appeared before'. The secret was revealed: 'Not a human figure visible, not a sign of human occupancy on that expanse! Was this the secret of her elation?' For the all-powerful, dreadful god inhabiting cities, 'was not here'. Hudson's godless vision of emptiness. His deathscape, outlined in *Land's End*, was of an 'illimitable wilderness', in keeping with the flat pampas of his childhood and the Downs.

Hudson chose to wander Savernake Forest as a place where he could be as far away as possible from his fellow beings in wild nature, so rare in England. The beech and oak trees, the dead leaves and nobody else about except birds led him to affirm that 'we are yet all of us at times hermits in heart, if not exactly wild men of the woods'. For real solitude, not the solitude of shutting yourself up in your room, which is a cage, a prison, is to keep at bay the miserable social conventions. You go to Savernake Forest to lose yourself and to find yourself invigorated. There you become free as you discover the forest's 'mysterious voice'. Hudson's technique is 'listening, thinking of nothing, simply living' so that he sat on a beech root and listened to a wood pigeon for half an hour in the 'profound stillness', without moving. He became the birdsong. Savernake was Hudson's transformative jungle. Out of Savernake emerged his bestselling *Green Mansions*, 1904.

Hudson never travelled in the Venezuelan jungle or up the Orinoco like Alexander von Humboldt. When he set his 1904 romance *Green Mansions* between British Guyana and Venezuela he was reclaiming territory as a South American fed up with the popularity of jungle romances set in Africa, headed by Rider Haggard. He also needed to establish somewhere remote, untouched from any city, where Stone Age tribes still existed, as his critique of contemporary society's mechanistic civilization. South America held undiscovered tribes in virgin jungle and still does. He had known corridors of sub-tropical woods along the River Plate coast, where tropical plants and animals had drifted down on tangled rafts of water plants called *camalotes* and colonized the shores. He had read the thorough Sir Everard im Thurn's *Among the Indians of Guiana* for background knowledge of jungle tribes. And he knew Savernake Forest.

Abel is a fugitive from the city. Like Hudson, his relationship with women 'excited no man's jealousy'; he loved children and wild creatures. He loathed material goods, politics, sports and the price of crystals. Hudson had described himself in *Hampshire Days*, written the year before in 1903 when he was sixty-two, as feeling estranged from his fellow men:

> When I look at them, at their pale civilized faces, their clothes, and hear them eagerly talking about things that do not concern me. They are out of my world – the real world. All that they value and seek and strain after all their lives long, their works and sports and pleasures, are the merest baubles . . .

In *Green Mansions* Hudson narrated his inner journey back to feeling alive and wild on the pampas as a child. Abel tried to imagine himself a simple Guyana savage 'with no knowledge of that artificial social state to which I had been born'. He does this by an effort of will: 'I emptied myself of my life experience'. This is an arduous transformation into a deeper instinctual self. Only then can Abel discover Rima, his anima or soul, a bird-woman.

Hudson could do the same only when free of others and alone in Savernake Forest. Here he bursts out: 'for this is the sweetest thing that solitude has for us, that we are free in it, and no convention holds us'. That was the attraction of being alone in wild nature; no social conventions. But he could not live there and so after each trip he returned to Paddington and again lost his freedom to regress to being a wild man of the woods. The deep nostalgia and melancholia behind the plot in this romance matched his own.

How to mutate into a savage when alone was akin to how he taught himself to observe. It's self-therapy manual territory. You must set off alone in order to properly concentrate. You must be mentally alert, with eyes, ears and nose picking up signs, observing. Hudson was very clear about suppressing thought as you cannot 'be in two places at once'. His task was 'to empty the mind as in crystal-gazing', for this allows you to recover 'a vanished experience or state of the primitive mind'. The mind becomes a polished mirror, 'undimmed by speculation' in which the outer, natural world becomes vividly reflected in inner reality.

Wherever he was, Hudson broke free of alienation to fuse emotionally with the world. His inner enemy was 'self-absorption' or that 'aloofness' that regards nature as outside subjectivity. This is Hamlet's famous lament about feeling 'sicklied o'er with the pale cast of thought'. That mental disease of weariness, flatness and staleness. The final analogy of this emptying of the mind to make it a mirror focuses on writing and memory. You sit at your desk and become what you are imagining and remembering as you write. You 'divest' yourself of yourself. You cry out like Smith, Hudson's alter ego in *A Crystal Age*, 'I am sick of thought – I hate it!'

This recovery of direct experience is the child's emotional view and writing was to keep as close as possible to that first shock of perception, for familiarity blinds and books dull. That is why Hudson was reluctant to return to the same places. The mind holds first impressions, which do not fade, to become 'treasures and best possessions'. In 1921 this vivid memory was enduring, defying time and loss. It was 'sacred'. And childhood memories are the primary ones.

I do not only mean that Jesuit boast of taking a child up to seven and making him a Jesuit for ever, or that rapid neurological development of the early years common to all of us, but also a writer's experience of the early shocks on the young mind when it trembles with fear and desire that then become stored and available. My early, first experiences were most vivid in Mauritius and in the terror of travelling in cargo ships and military planes. These experiences form the core of recurrant dreams. Hudson recalled these early images out of the solitude of writing and recreated them as little shocks of pleasure in his prose.

In June 1908 Hudson published an essay in the *Egotist*. He was on top of the world and considered a modernist. This essay is formally a whole, with slight avant-garde flourishes. It opens with an abrupt 'That American from Indiana' and mocks his fellow countryman's attitude to the past. 'Poor Indiana, that once had wildness and romance and memories of a vanished race', he wrote, 'and has only its pretty meaningless name.' Then he introduced his theme. When a boy on the pampas, he came across a print of Stonehenge in a folio titled *A New System of Geography* by A. F. Busching. He was so impressed that he mischievously cut it out to keep. This 'half-shattered skeleton of a giant' on plains he thought were like his childhood pampas lured him to England. Stonehenge almost reached the sky in his childish mind, as do all things seen in childhood. An Italian peasant he'd met in Buenos Aires told him of a childhood vision of a giant heron that had landed during its migration and the whole village had gone to see this giant bird.

But when he finally got to see Stonehenge, he found a cluster of poor grey rocks 'looking in the distance like a small flock of sheep'. He walked around the ruins, checking sparrows' nests. He returned years later, before dawn, to witness a solstice. In the dark, he heard an animal din of cows and dawn birds that 'expressed the mystery and glory of the morning'. For Stonehenge, a Druidic sun temple on plains taken over by the military, stirred up a sense of the depth of time, a transitoriness of all things human. Later he attended Sunday service at Shrewton church – he always attended church in villages – and wished that Stonehenge could be taken as seriously as Salisbury Cathedral.

In a letter to Garnett, he said that Stonehenge 'appeals to me more than any place builded by man'. On the frontispiece of the first edition of *A Shepherd's Life* was a drawing of Stonehenge by Bernard Gotch, without people. He was critical of the noisy crowds celebrating dawn. My first visit to Stonehenge was also without barriers. I wandered about the blocks, meditating on ruins. Today, tourists are railed off from the stones. You cannot stroke the Welsh bluestones.

At this same post-naturalization time, Hudson published an illustrated children's book, *A Little Boy Lost*, its title lifted from a William Blake poem in *Songs of Innocence and Experience*. An immigrant boy from Southampton escapes home and wanders the pampas (reversing Hudson's journey), meeting Indians, rheas, an Alice in Wonderland Queen of Hearts and wild horses, in search of his mother. It turns out to have been a dream. However, it was a trial for his later memoirs *Far Away and Long Ago* with many overlaps, down to finding wild potatoes. But Hudson rejected this fantasy journey into wilderness for its lived version. He sought to explore 'little nature thrills', with a boy's half-frightened, half-fascinated sensibility, a 'feeling I myself experienced when out of sight and sound of my fellows on the great level plain'. He added in legends 'heard from my gaucho comrades'. On 6 January 1920, he received an advance of $150 for the US edition, with new illustrations by Miss Lathrop. But as we know, this newfound money meant nothing to him.

A personal version of *Little Boy Lost* appeared in *The Land's End*, 1908, when Hudson, alone on the black granite cliffs in winter, facing crashing Atlantic waves, remembered the poem 'The Hunter's Vision' (by William Cullen Bryant, the poet on his tombstone) that he first read as a little boy on the pampas. He said: 'I read it first in my early years, and though it was poor poetry it powerfully affected me' (a study could be written about the influence of 'poor poetry' on budding poets) as he identified with the hunter's vision of his childhood home. Hudson was 'divided by death and change and absence from his own kin who were dearer than all the world to him'. Utter loneliness in 1907 was due to 'the impassable ocean of death [that] separated me from my own people' (meaning the Atlantic and the long sea-passage). The fantasy, as he fell asleep, was seeing the 'old roof and all those I first knew and loved on the earth', then he leapt up, wide awake, and resumed watching gulls.

Chapter 14

The Great War and Hudson's Death

The Great War doesn't seem to have affected Hudson, at least not in his prose rambles. Far too old to enlist, he viewed the war as a longed-for purge of an over-comfortable society. Not a mention, then, of the Great War in his books, despite surviving twelve bombing raids as he boasted to Alfred Knopf, watching Zeppelins and writing *Far Away and Long Ago* during such 'dreadful' years. Violet Hunt voiced reality when she wrote that 'we talked in those days of war and nothing but war', and Mrs Lemon thought the Great War 'worried him tremendously'.

In a 1915 letter to Garnett, he linked the war with what he loathed about the urban way of life:

> You think it a 'cursed' war, I think it a blessed war. And it was quite time we had one for our purification and our [word missing] from the degeneration, the rottenness which comes

of everlasting peace . . . the blood that is being shed will purge
us of many hateful qualities – of our caste feeling, of our
detestable partisanship, our gross selfishness and a hundred
more. Let us thank the gods for a Wilhelm and a whole
nation insane with hatred of England to restore us to health.

When it came to war he remained an outsider to English patriotism. He would have agreed with Ezra Pound's lines about a
'botched civilization' where 'old men's lies' and that 'old bitch gone
in the teeth' (Queen Victoria) still ruled. And that despite knowing
Lord Grey and losing his close friend Edward Thomas in Arras,
France. He reacted against the herd mentality and shared William
Morris's claim that the leading passion of his life had been hatred of
the way modern civilization stifled a healthy alternative. But how to
express that hatred took very different paths. Morris became a 'practical socialist' and Hudson worked tirelessly for the RSPB. For both,
hard work was pleasure. In a letter to Roberts in 1919, Hudson
lamented that Morris wasn't scientific. Instead, his worship of earth
and rain and harvest work as a vision of happiness struck him.
Morris put perfection in art below the healthy life. Hudson added:
'I put him far above the great artists of his day.'

In fact, 1914 was the year when Hudson became a successful
author. Richard Aldington, the twenty-two-year-old literary editor
of the *Egoist*, and later war poet and prolific self-exiled writer,
assessed Hudson's literary reputation in the issue of 15 May 1914.
This was just a month before anarchist Gavrilo Princip's assassination of Archduke Franz Ferdinand would kick-start the war.
Aldington might not have met Hudson, but tackled his reputation
as the best modern prose writer in English.

In a mixed survey, he initially voiced disappointment. He had
started with *A Crystal Age*, what he called a 'vaguely Morris-like

Utopian' work, the worst specimen possible: 'Its damnable Sunday-school ethics were paralysing'; he wanted to run out into a street, shake hands with a navvy and thank God for 'filth, beer, noise and naturalness'. The story 'El Ombú' he found fresh and vivid. Hudson was a poet writing in prose (John Masefield made the same point, as did Ezra Pound in 1920); moving to *The Purple Land* he praised Hudson's reconstruction 'of a life that none of us Europeans has ever lived – the life of the gaucho and ranch-keepers in that primitive land'. *Green Mansions* had Aldington comparing Hudson to Conrad. He then dismissed the nature books and ended with a plea to the very much alive Hudson, who read it: 'I wish Mr Hudson would write some more tales and leave grasshoppers and cirl buntings.' A reader wrote to the editor to complain of this 'insulting' article.

What's fascinating at this moment in time, May 1914, is that Hudson, promoted by Ford, mingled with James Joyce and others in this magazine as a modernist. He also published in the *English Review* with Ezra Pound, a magazine that Aldington considered to be England's best literary journal. The war changed all that.

One of the last books Hudson published after the war ended was his revised *Birds of La Plata*. It's intriguing that in 1920, aged seventy-nine, he reverted to Argentine birds. Marcos Victoria, Argentinian scientist, thought that this book was 'the best lesson of love that Hudson left to Argentina'. Many of the birds' Latin species names have changed. The page proofs are held in the British Library, with Hudson's scribbled notes and additions, but I found little of personal interest. H. Gronvold's new illustrations are now kept in the Asociación Ornitológica in Buenos Aires. Only in the Argentine translation are Hudson's facts checked, updated and corrected. For example, in 1920 Hudson listed four species of swallow, now there are twelve; his four woodpeckers have risen to forty-one. In his book, Hudson recalled 192 bird species, but by 1972 some 1,146 had been described in Argentina. The Latin species names were also updated by Herminia and José Santos Gollan in the 1970s.

To test how vivid his memory of Argentine birds was in 1889, Hudson made a list of species observed in La Plata and Patagonia. Then he added his impressions, both visual and auditory. Then he revised this list, added more from memory, re-read his Félix de Azara, and put it into a drawer so that his 'subliminal subconscious' could work further on it. He had listed 226 species, some sixteen to eighteen more species than in Britain. Of the 226, his memory could not recall some ten and he had completely deleted one species. That left him with 215. Running through birdsongs in his head, he had forgotten the song of thirty-four species, but could replay in his memory the rest: 154 birdsongs.

This amazing feat of recall confirms how deeply emotional birdwatching had been for him, that he had laid down aural memories that he could replay twenty-five years later and, for the second edition of 1920, forty-six years later.

Just as Hudson recalled birds from his past for the second edition in 1920, he also, aged eighty-one, published an uncollected piece, 'Do Cats Think?', which summarized a life-time of guessing what animals think. It was included in the *Cornhill Magazine* in 1921, a popular magazine, founded in 1860 and first edited by William Thackeray, that was so inoffensive that it even published writing by Queen Victoria.

Hudson's article is written from near Falmouth, standing at his lodging's window, observing the antics of a cat, 'this largest-brained and most perfect mammalian'. Hudson's age kept him indoors, but he didn't stop observing life around him. One cat reminded him of another one, which also revealed how quickly cats could learn, that cats had 'ideas' and proved that they could reflect. He experiments by throwing salt or sugar on this cat, or by keeping a door closed. He has persevered with his 'little incidents' tradition. At one time, he translates into words what the cat thinks: 'How many times have I explained to you that the door must not be shut tight . . . Are you so hopelessly lacking in intelligence that you cannot understand it?' He dismisses the view that animals are dumb; a cat is capable of thought.

Then he tells an anecdote he overheard – as he once told stories recited to him by gauchos – concerning a cat instructing a dog to carry her young to another room. Finally, Hudson tells, verbatim from her written account, a tale about a bishop's wife's telepathic message from her cat. Hudson's curiosity and speculations never dimmed. His prose is serene – the writer alert – and not a whine about health, the year his wife died. And like any bird-lover he'd campaigned against cats. A writer going by the initials C. S. E. met him through Alfred Knopf in 1921 to find him sombre and brooding, but his face lit up suddenly when struck by some comment. He was simple and unpretentious, and graced by shrewd observations. Here was the observer of cats.

Hudson was very modest, as his title *A Traveller in Little Things*, 1921, implied. He often told other people's stories, whether from a gaucho or one of his many correspondents. Little problems 'vexed my mind for years' as he tried his own improvised experiments out-of-doors. He was aware of his 'meagre observations and comments' – the way they were related counted more, so he laboured at style. He was out of date before he started writing, a field naturalist deal-ing with the 'life and conversation of animals'. He could empty himself in his listening, rather than boast of his conquests.

Any little event could become book-matter, like an encounter with a bat in a deep lane near Selborne. Here he tested the bat's aerial skills by swishing his walking stick round and round his head and never once hitting the bat. He did this fifteen or sixteen times, so that 'these bats in Selborne lane taught me more than all the books'. He recalled bats in Argentina while sitting round a campsite with nine gauchos 200 miles south of Buenos Aires, near Sierra de la Ventana. He'd picked a bat up, sealed its eyes and freed it in a room with hanging rope, but it flew about without once touching the walls. Such little problems like this intrigued him until his last days.

Any 'very little thing', such as the scent of a flower or the cry of a bird, reminded him of some encounter. He remembered his first view as a child of a huge circle of mushrooms or a sensitive plant, struck with a 'sense of mystery'. Hudson, a traveller in little things, found the great mythical experiences in little incidents that nobody else recorded but reminded him of something else.

In my circling round the man W. H. Hudson, I've resorted to monuments, to his books and articles, to his friends' insights and to letters and reviews as I slowly generated what I call a cubist por-trait of the pampas-born man. I'm convinced that he didn't want to be understood, that he enjoyed his deep privacy and obviously didn't want any biographies to be written. And yet I learned about the man and how an observer reveals himself through what he observes. But he wasn't solipsistic or a monad. His world was not limited to his subjectivity. He broke through to the other lives in nature, he learned bird talk and cat behaviour. He believed all life shared this earth. He was a true animist. His thinking through anecdote, rather than conceptual thought, made him appear simple. But I am left with an image of an unknowable and complex man.

Like Darwin, Hudson recorded few dreams, but was fascinated by the 'unexplored wilderness of the mind'. In a letter to the poet William Canton he described a nightmare, with a storm, whirlwinds, waterspouts and a huge black spot. Nobody listened to Hudson as he tried to explain . . . Hudson did not interpret his dream but what stood out was the dreamer's loneliness. That nobody listened to him is an apt description for Hudson at that time.

I could only find one other recorded dream, which occurred when he was eighteen. He stood on the plain and saw a dark object hurtling towards the earth, with huge iron bars. He immediately thought this was the end of everything. Then came the crash. Waking up, he realized that it had been a thunderclap. As if a dream jumps backwards to create a forward narrative from a sensation of noise.

However, Hudson avoided introspection, and instead scrutinized the material world. As a bird lover, he envied the freedom of flight, especially in *chajás*, the great crested screamers. He longed for wings, but refused to fly in air balloons or airships, though twice succeeded in what he called levitation dreams, floating above the earth without effort.

It's pointless pigeon-holing Hudson's writing into bland categories like nature sketch or romance, and hard to define his public persona or his politics. In 1921 he wanted to be seen as an 'unclassified man'. In the public sphere, he busily defended the uniqueness of birds in letters, pamphlets, and attending and chairing meetings of the RSPB. He corresponded with Keir Hardie, admired Edward Carpenter and Tagore, was appreciated by ex-US President Theodore Roosevelt, British Prime Minister Stanley Baldwin and Foreign Minister Edward Grey. He closely followed the suffragette debates, the Irish question and signed petitions on public issues. Yet he remained deeply private and excessively individualistic.

The primary root of his individualism was his Puritan upbringing with that emphasis on individual conscience and direct pleading with God. He learned that emotional self-control and introspection were crucial. He rejected sloth and idleness. He always respected hard work. In fact, Benjamin Franklin's thirteen 'Virtues with their Precepts', including Temperance, Frugality, Sincerity and Moderation, would fit Hudson. My grandfather handwrote Franklin's virtues for his son, and my father handed the letter down to me. Being American heretics in a recently independent Catholic Argentina increased the Hudson family's isolation. His family were unique, literate, kind, curious. I think that Hudson was actually 'happy' until his teenage illness hit him. For Borges the most memorable line in all the literature he'd read came from how Hudson's Richard Lamb interrupted the study of metaphysics because he was plain happy.

There was no civic society in Argentina during his childhood years (it's still weak), no civic co-operation. There were tyrants like Rosas and Urquiza or demagogues like Mitre and Sarmiento. Living as a family in the country fuelled individualism against the state, epitomized by the roaming gauchos, who rejected laws and taxes and suffered marginalization. Perhaps this combination of puritanism, exile, absence of civic values and gaucho freedom paved the way for Hudson's intense privacy. Then, his long exile in London, only attaining English status late in life, simply confirmed his individualism. He always felt 'poisoned by contact with the crowd-mind'. After all, few exiles or immigrants are integrated into the public arena and assume public responsibility. My family were a typical case of non-involvement. Though they read the papers, they never joined a party, nor voted. My father argued that artists shouldn't take sides, and remain independent. But in fact defended the status quo. Against their lack of communal civic values, I've always voted in protest, usually for the Greens, and don't own a car, but have rarely demonstrated or marched. Perhaps I too am reluctant to join a group, play in a team game and think collectively. I'd read my Marcuse about thinking non-politically being a political position and I edged around the Marxists at university, but somehow English concerns were not mine. Neither the flag, nor the monarchy, nor the public schools, nor the stately homes – the list is longer – meant much to me. I realise now that this lack of commitment was the consequence of my colonial heritage.

From this complex outsider status (puritan, exile as an American in Argentina, exile as an Argentine in England), Hudson defined himself against many elements of English sensibility, like fox hunting, beef eating in its Hogarthian excess, having dogs like Pekinese as pets, the upper classes, the London literary world, dry-fly fishing, rhododendrons, pheasant breeders, Kew Gardens, Christmas carols, caged birds, game-keepers, Methodism, bird-egg collectors, flower gardens, all sports, Italians, motorcyclists, even the idea of 'home' . . . the list of what Massingham called his 'salty prejudices' is endless. I'm sure his talk constantly reflected such 'petty irritabilities', as a *Times Literary Supplement* reviewer noted in 1902.

He grasped a fundamental truth about how contact with nature can heal us. His individuality is based on him knowing that every moment in any life is unique and transient. He many times asserted that he never returned to a place where he'd been happy. As if place

Violet Hunt.

and subjectivity were determined by Heraclitean flux: you can't be happy twice. And then there's gaucho freedom to roam on a horse, not pay taxes or join the army. Behind all Hudson's opinions and attitudes, there's his quest for freedom from societal demands. Privacy is just another kind of freedom. Hudson was a libertarian, in favour of minimal constraints on an individual's freedom of action.

Freedom was also manifested in how he wrote. He congregated sketches because they paid – he labelled himself a journalist – and then turned them into episodic books, each chapter almost autonomous, and was paid a second time. Virginia Woolf noted his fragmentary brilliance, reflecting his interesting mind (and regretted not meeting him). The unifying thread in all he wrote was Hudson himself, surfacing through anecdotes and memories as voice. So the writer is the man talking to his reader (and hiding the sweat to convey this tone). His readers' responses were his form of being integrated back into society. Readers like Philip Gosse and Edmund Blunden, and many others, devoured his books in the trenches.

Hudson had allowed them to re-create an idealised England in the midst of the mud, stench and dying.

I envisioned his quest as struggling to belong to all life as knots in a vast, living, natural network. He could never be alone and was protected by his birds and flowers and memories of gauchos and horses. That he shared companionship with his dead mother and with his wife Emily and quickly followed her to death, that he fell in passionate love with Linda, that he adored his younger sister Mary Ellen and brother Albert Merriam, that he met, listened and chatted with countless people from peasants to poets contradicts being a recluse.

His canny friend Edward Garnett, in a 1903 essay, placed Hudson between science and poetry because he 'refused to divide man's life off from nature's life'; because Hudson's understanding of sentient life involved feeling, imagination and aesthetics. So I turn to William Rothenstein who placed Hudson in 'a class alone'. Most of his friends voiced what Massingham wrote in 1923; he was 'difficult to classify' because he owed nothing at all to contemporary fashions and tendencies in literature.

Hudson's alter ego Abel affirms at the end of *Green Mansions*:

> That is my philosophy still: prayers, austerities, good works
> – they avail nothing, and there is no intercession and outside
> of the soul there is no forgiveness in heaven or earth for sin.
> Nevertheless, there is a way, which every soul can find for
> itself . . . In that way I have walked, and self-forgiven and
> self-absolved . . .

Every person must work out their own salvation; that's their freedom, rather than adopting another's system or thoughts.

In the Worthing cemetery records, only the actual date of Hudson's burial – 22 August 1922 – is recorded. His housekeeper,

Walter de la Mare.

Mrs Jessie McDougall, said in a letter that 'we were very disappointed to see few people there' at the funeral. But Morley Roberts was there with friends Violet Hunt, Alice Rothenstein, Edward Garnett and publishers J. M. Dent and George Duckworth. Linda Gardiner had sent flowers with a message to her best and most beloved friend. When he died on his own in his St Luke's Road tower, the pretty, 'Rosetti' housekeeper (Violet Hunt's description of Jessie) had called his friend of forty-three years, Morley Roberts, to arrange the body and fold his arms on his chest. He remembered 'in death he had the muscles and thews of some great athlete'.

The best account of his last days is that of Violet Hunt. She evoked him as a 'graceful, perfect old man of unknown age', who had steel-like limbs and cared about his clothes. He remained agile and upright until the end, even jumping on the no. 31 bus at Sheffield Terrace, to her shock. They had talked about the afterlife and Hudson, in his tenuous, wistful voice, had asked her to read his essay 'The Return of the Chiff-chaff', as what he thought about death was there. She went to see his dead body on the first floor of 40 St Luke's Road. He had complained of indigestion and chest oppression. His doctor had ordered Bovril and weak tea. Hudson was fond of tea, she recalled. He had said goodbye, not goodnight, to his housekeeper and then died on his own as he had wanted, like a horse or a cat. By his bed on a green baize table, he'd left his writing pad and pen in case something occurred to him. He was a writer to the last night.

In 'The Return of the Chiff-chaff' Hudson evoked a pond, some alders and himself alone, watching chiff-chaffs and grieving his recently deceased wife Emily (without naming her). Instead of his usual skill of 'divesting himself of himself' in the wild, he suffered a 'monstrous betrayal'. Yet he learned from the birds how to live in the present. They have no knowledge of death. Birds do not ask after meaning. They just are. Grief was transformed when he observed the common things and listened to birdsong, as if the dead loved ones have melted back into this visible world, adding 'tenderness and grace and beauty not its own'. That is, all the past dead, led by his mother, invisibly accompanied him while rambling in the shires, as if he walked always with ghosts.

I can now understand Hudson's notion of memory as being sacred. Walter de la Mare, an acquaintance, acutely described this double reality as being haunted, as if behind Hudson's descriptive prose 'a profoundly concealed innate voice of memory, rather than a novel earthly experience, were struggling to express a secret knowledge'. Hudson had faith in memory, which worked emotionally and accompanied him while alone in nature, or, better, in his study writing about nature. His memories were unique to him, so deeply private. But this simultaneous experience of outer reality and inner, ghost-like memories is too subjective to become explicit and doesn't survive death. Hudson can only prompt a reader to react against passing time and the monstrous betrayal of death in a similar way. You need to be alone to sense your loved ones and let memory roam.

Such was Hudson's reputation on his death that *The Times* referred to him in its leader and published an obituary on 19 August 1922. Alluding to his 'Virgilian sensibility', it said he was 'unsurpassed as an English writer on nature'. He was placed with Wallace and Bates as the 'greatest English writers on South American natural history'. Twice emphasizing his 'Englishness', yet underlining his South American works, would have pleased Hudson. But he wasn't one of them, for he was a native Argentine.

In Hudson's will, written while residing at 23 North Parade, Penzance, where he wintered, he asked for all his notes and letters to 'be destroyed by my Trustees' and left money to close friends Linda Gardiner, Morley Roberts and Edward Garnett, He also left £25 to his housekeeper, who could continue to live in the basement at St Luke's Road. Nothing was left for any Argentine family members. By the time of his death he'd outlived his siblings and considered himself English. The bulk of his estate went to the RSPB (some £5,800; around £297,000 today). The RSPB sold his library on 13 July 1923. The booklet describing his collection included two childhood favourites, his 1820 Bloomfield, *The Farmer's Boy*, and his 1797 Hurdis, *The Village Curate*. The RSPB raised £348.16 (worth today some £17,000).

His life had illustrated Kipling's poem 'The English Flag': 'And what should they know of England who only England know?' Kipling's Indian years, similar to Hudson's Argentine ones, gave him, in T. S. Eliot's words, a 'universal foreignness', his persona in his works came from a 'remoteness as of an alarmingly intelligent visitor from another planet'. Does that capture Hudson's position?

There's no doubt that Hudson's English reputation was at its highest when he died in 1922. A leader and an obituary in *The Times*, a monument in Hyde Park opened by the Prime Minister, best-sellerdom in the States with *Green Mansions*, Hudson was much loved by many admirers and readers. He has been in print ever since. His work now forms part of the ecological movement of protest against animal cruelty, though his legacy is literary.

When the initial meeting was held on 27 November 1925, at the publisher J. M. Dent, to organize a memorial to W. H. Hudson, Cunninghame Graham proposed a bust for the National Portrait Gallery: 'It is incumbent on us as Englishmen to honour the memory of one of the greatest English writers who ever held a pen.' But Philip Gosse remembered that his father Edmund Gosse (a friend to Hudson) found this admiration for Hudson 'out of all reason'. After Hudson had been compared to Rubén Darío, the great Nicaraguan-born *modernista* poet, and been called a genius, Gosse Sr, sole dissenter, had insisted that Hudson's reputation should not be so exaggerated. Philip Gosse guessed later, in 1937, that Hudson's posthumous reputation would not 'hold quite so high a place'. He was right.

In Argentina, meanwhile, Hudson was rediscovered in the late 1930s through Dr Pozzo, and taken up as the grandest gaucho writer by Ezequiel Martínez Estrada and then by Jorge Luis Borges in a famous little essay of 1941. But Hudson's work was more than a *gringo* version of Argentinian reality, for he was also considered a foundational ornithologist with his *Birds of La Plata*. He remained a scientist, one of the earliest to record natural history in the River Plate. The Museo Hudson in Quilmes safeguards this reputation. That's his *'destino sudamericano'*, in Borges's words about his grandfather's death.

I mentioned the American reputation of Hudson being based on his romance, *Green Mansions*. This status reached a high point with a Hollywood film. In 1921 Hudson worried about selling the film rights

Sketch of W. H. Hudson by A. D. McCormick in
his final year.

to *Green Mansions*. A year later, according to correspondence with
Knopf, he earned $2,500. But it was only in 1959 that Mel Ferrer cast
his wife Audrey Hepburn as Rima, opposite Anthony Perkins's Abel.
Filmed in Panavision, with a score from the Brazilian composer
Heitor Villa-Lobos, it was a box-office flop. The opening credits with
Green Mansions in green over a Venezuelan waterfall promised
authenticity. Ferrer's team spent three months filming background
shots and brought plants, birds and animals back on a ship to film the
rest in California, but Ferrer, like Hudson, had never stepped into
Venezuelan jungle. The scriptwriter – Dorothy Kingsley – altered the
love story, added a dash of violent Latin American politics and a happy
ending, where Perkins and Hepburn vanish into the jungle together,
rather than Rima being burnt to death and Abel suffering his intense
grief. Some critics thought that Audrey Hepburn was no match for
Rima. She was twenty-nine years old and Rima was a girl of seventeen
and tiny at four-foot-five, but I disagree. Audrey Hepburn, with her
elfin figure, was far more a Rima than Epstein's sculpture.

Over the years W. H. Hudson has become what Ford Madox Ford called a 'healer'. In my case, he accompanied me in countless country walks. His skilled, attentive observing released me from my over-bookish, city self to glimpse something more alive than the Ordnance Survey maps and botanical and bird guide books in my back pockets. The persona that surfaced from his words was never comforting. His quirkiness – manifested in lucid prose – both irritated and moved me. In so many ways, he would have disliked my urban modernity and taunted me as an enemy (in Baudelaire's sense).

However, he was from an almost sealed-off past and it was a real struggle to try to get under his skin or elucidate how he wrote. As Joseph Conrad had found. I was not alone in being puzzled by the man himself.

Close, literary friend Morley Roberts reiterated that Hudson had kept his soul in a 'strong, secret place'. A *Times Literary Supplement* reviewer in 1902 asserted that 'Mr Hudson . . . is nothing if not unlike other people', but without assessing why. A woman friend repeated a similar insight: 'No one knows him really and I believe he is proud of it. It pleases him to think how well he's kept his secret.' But is there a secret?

All witnesses to his life insisted on Hudson's hardly ever baring his soul, just as his skilled writing was 'unanalysable', in Richard Curle's phrase. There are enormous gaps in his thirty-two Argentine years and in his English life until, in 1916 with the Knopf edition of *Green Mansions*, he managed to live from writing when already an old man (it had sold 10,000 copies by 1917). That he burned his notes, two books of poems and some 2,000 letters, in his own calculation, when near death suggests a desire for oblivion and a gloating over the frustrations of future biographers. But confronting this mysterious, quasi-mystic misfit unravelled some of the threads of my uprooted life.

Hudson's exemplary work with the RSPB leads to Rachel Carson's predictions of doom (she opens with the death of bird-song), climate change activists, anti-badger cullers, Greenpeace anti-whalers and so many more animal protest movements. He can be read with Thoreau, with Gandhi, with D. H. Lawrence. His motto, which I borrow from Linda Gardiner, would be 'DON'T KILL'. He was a genuine animist.

My aim has been to bring Hudson to life, but not explain him. He himself insisted in a letter to Edward Garnett that 'a man is so much better than his books', as if only a minute part of a myriad self enters the written works. He had, of course, always been a writer about to be discovered, who failed to find the form for his visions. He had to earn to survive in England, rather than compose high art. And didn't care that much about posterity. But he has become, in poet P. J. Kavanagh's words, 'almost a cult'.

Afterword

You never finish a work, you abandon it, wrote Paul Valéry. And especially so with biography, where new facts and perceptions keep turning up. Here are two examples. A key photo of the Hudson plaque on his home at 40 St Luke's Road was too obscure to read. I'd tried to snap it several times, but the light was never right. One day I was there with Andrea and the sun dipped out of clouds and shone brightly on the plaque, making it suddenly easily readable. I took the picture. We were standing on the pavement looking up at the house when Andrea noticed a woman entering the main door with a key, and she turned and stared at us. Minutes later, from the top window of the tower, a man leaned out and shouted 'Do you want to come up?'

Without a thought I answered, 'Yes, we do'. We were let into a hall with tiled mosaics and a circular staircase going up three floors. 'These must be the original tiles', said Andrea. We walked up as there was no lift. At the top there was a bit more of a staircase, and the man appeared at a door. He was Paul Polydorou, a photographer. He showed us around his top floor, attic flat, with his stunning, framed photos of Syrian ruins and desert tribe people spread round the drawing room. He'd just had an exhibition. He had let in other

Hudson fans. He'd even read some of his work. When he showed us the tower room, now a bathroom, I realised that this was Hudson's writing space. A gust of cold air hit me and made it suddenly real (as well as a raven sitting on the ledge outside). Hudson had called it his 'Illimani' it was so cold, like an Andean peak. There was no heating in his time. He sat with a rug over his feet and carefully wrote from his notes. Later, on leaving the top flat, I noticed a manhole cover with the date 1862. In Hudson's time, it was a new housing development.

A second example came with Enrique Pedrotti, who wanted me to meet a descendant of Hudson's neighbour Juan Davidson, owner of an enormous *estancia* next to the Hudson's in Quilmes. He was Viscount Davidson, whose great-grandfather had bought his estancia in 1843 and then the land around Hudson's hut, including the hut itself in 1877, from Hudson's elder brother Daniel. His grandchild had donated the hut, with four hectares, as a museum, in 1949. I'd bumped into Viscount Davidson by chance in The Travellers Club, where I was a guest. He insisted on taking me out a few days later to the Canning Club, in Lady Astor's Inigo Jones designed house. Viscount Davidson was the son of the co-founder of Canning House, which had an offbeat library I often used from the 1960s in an imposing house at 2 Belgrave Square. It was a meeting place for Latin America and Britain. I'd given talks and run conferences there, despite finding the place a bit stuffy. I turned up at the club, named after the foreign secretary George Canning, who was also, briefly, prime minister. We sat in the empty library as Viscount Davidson pulled out map after map, dealing with his family's long-running court case disputing Juan Domingo Perón's expropriation of their land in 1949. In the end the family got no compensation. I had toured the ex-Jesuit, then Dominican, monastery – the original *estancia*, as I've recounted. Davidson was full of stories and hearsay. Inspecting the maps, including one published especially by Stanford's as an out-folding book, I could establish that Juan or John Davidson had finally bought out the Hudsons in 1877. But I couldn't get any further into Hudson's relations with his famous radiologist son, who was his knighted doctor. For some taboo reason, they couldn't talk about growing up on the pampas.

Finally, I had managed to contact Sr Gayol at the Burmeister archives in the grand Museo Argentino de Ciencias Naturales Bernardino Rivadavia. The reason for not being able to meet him earlier, I discovered, was that he was recovering from a stroke. Only he had the keys to open the cupboards in his room. As he was fiddling with a bunch of keys, this gentle man told me that recently someone from above had decided to junk two hundred years of useless documents and build a computer room instead. He'd had to hurry back from his first holiday for four years to resolve the crisis. I said that's happening in London, too. He also told me that deadly black widow spiders were colonising the books, but that was maybe a scare story. Quite a lot of Burmeister's correspondence had been digitalised, but not between 1865 and 1871. So I went through folder after folder, saw his complaints at not being paid, lists of expenses, letters from well-known people, but nothing on Hudson. Maybe Hudson was too young, too much an amateur?

Gayol took me to meet the Museo's director, an ornithologist called Dr Pablo Tubaro, who confirmed that Hudson, who wrote the first book on Argentine birds, is still consulted as an ornithologist. But my visit wasn't fruitless, as in the hall leading to the staff rooms, there was a large photo of Hudson among scientists such as Jurado, Biraben and Burmeister. It was Burmeister's solo effort to expand the 1812 Museo, on another site near the Plaza de Mayo, that led to the building of the modernist and eclectic Museo on the circular Plaza Centenario from 1925 to when it was finished in 1940.

Burmeister is buried in the crypt. He came to a sorry end. He'd stepped on to a chair to open a blind as it was so hot, slipped and crashed into a glass case that shattered and killed him.

Of course, more letters will turn up. Maybe something survives concerning Abel Pardo, a Radical in politics and his first translator in Argentina, but I couldn't stay long enough in La Plata. Or with Ernest Gibson at his *estancia* Los Yngleses, a fellow ornithologist. Hudson's name must figure in more lists of soldiers on the frontier and of taxes paid. More will be eked out concerning Hudson's thirty-two years in and around Buenos Aires. And there must be photos somewhere of Emily his wife and Linda Gardiner, his platonic lover, in their prime. But for now I must abandon Hudson.

Bibliography

(London unless stated)

W. H. Hudson editions

Adventures among Birds (Hutchinson, 1913).

Adventures among Birds, introduction by Robert Macfarlane (Collins, 2013).

Afoot in England (J. M. Dent, 1924).

Aves del Plata, prologue by Dr Marcos Victoria, tr. Herminia Mangonnet de Gollan and José Santos Gollan (Buenos Aires: Libros de Hispanoamericana, 1974).

Birds and Green Places: A Selection from the Writings of W. H. Hudson, selected by P. E. Brown and P. H. T. Harley, introduction by P. H. T. Harley (RSPB, 1964).

Birds and Man (Duckworth, 1930).

Birds in London (Duckworth, 1928).

Birds in Town & Village (J. M. Dent, 1919).

Birds of a Feather: Unpublished Letters of W. H. Hudson, ed. with an introduction by Dennis Shrubsall (Stratford-upon-Avon: Moonrakers Press, 1981).

Birds of La Plata (J. M. Dent, 1920).

Birds of La Plata, introduction by Richard Curle and colour plates by S. Magno (Penguin, 1952).

The Book of a Naturalist (Hodder & Stoughton, 1919).

The Book of a Naturalist (Hodder & Stoughton, 1923).

British Birds, with a chapter on Structure and Classification by Frank E. Beddard (Longmans, Green and Co., 1897).

A Crystal Age (New York: Doric Books, 1950).

Dead Man's Plack, an Old Thorn & Poems (J. M. Dent, 1924).

Diary Concerning His Voyage on the Ebro *from 1st April to 3rd May, Written to his brother Albert Merriam*, with notes by Dr Jorge Casares (Hanover, NH: Westholm Publications, 1958).

'Do Cats think?', *Cornhill Magazine* (May 1921).

Far Away and Long Ago, introduction by Robert Cunninghame Graham (Buenos Aires: Guillermo Kraft, 1943).

Far Away and Long Ago, afterword by Nicholas Shakespeare (Eland, 1982).

Far Away and Long Ago: A History of My Early Life (J. M. Dent, 1918).

'Foreword', Edward Thomas, *Cloud Castle and Other Papers* (Duckworth, 1922).

'Goldfinches at Ryme Intrinsica', *English Review*, 2: 6 (1908), 246–54.

Green Mansions (Duckworth, 1922).

Green Mansions, introduction by Carlos Baker (New York: Bantam Books, 1965).

Green Mansions, ed. with an introduction and notes by Ian Duncan (Oxford: Oxford University Press, 1998).

Guillermo Enrique Hudson, Vida y obra. Bibliografía. Antología (New York: Hispanic Institute, 1946).

Hampshire Days (Longmans, Green and Co., 1903).

A Hind in Richmond Park (J. M. Dent, 1929).

Idle Days in Patagonia (J. M. Dent, 1924).

The Illustrated Shepherd's Life, foreword by P. J. Kavanagh (Bodley Head, 1987).

Lands' End (Hutchinson, 1908).

Landscapes and Literati: Unpublished Letters of W. H. Hudson and George Gissing, ed. with an introduction by Dennis Shrubsall and Pierre Coustillas (Salisbury: Michael Russell, 1985).

Letters on the Ornithology of Buenos Ayres, ed. David R. Dewar (Ithaca, NY: Cornell University Press, 1951).

A Little Boy Lost (Duckworth, 1905).

'A Memory of the Ancient Time', *English Review*, 11 (June 1912), 349–65.

Men, Books and Birds, with notes, some letters and an introduction by Morley Roberts (Eveleigh Nash and Grayson, 1925).

The Naturalist in La Plata (J. M. Dent, 1939).

Nature in Downland (J. M. Dent, 1923).

Obituary for Lady Dorothy Grey, *Speaker* (3 March 1906).

153 Letters from W. H. Hudson, ed. with an introduction and explanatory notes by Edward Garnett (Nonesuch Press, 1923).

The Purple Land, illustrated by Keith Henderson (Duckworth, 1929).

The Purple Land that England Lost (Sampson Low, Marston, Searle and Rivington, 1885).

Rare Vanishing & Lost British Birds, compiled from notes by Linda Gardiner with twenty-five colour plates by H. Gronvold (J. M. Dent, 1923).

A Shepherd's Life, introduced by Adam Thorpe and illustrated by Howard Phipps Little (Wimborne Minster: Little Toller Books, 2010).

A Shepherd's Life: Impressions of the South Wiltshire Downs (J. M. Dent, 1923).

'So Far', *Gentleman's Magazine*, 261 (November 1886), 502–51.

South American Sketches (Duckworth, 1909).

Tales of the Pampas (New York: Alfred A. Knopf, 1916).

La tierra purpúrea & Allá lejos y hace tiempo, prologue and chronology by Jean Franco (Caracas: Biblioteca Ayacucho, 1980).

A Traveller in Little Things (J. M. Dent, 1921).

The Unpublished Letters of W. H. Hudson, the First Environmentalist 1841–1922, vols 1 and 2, collected, transcribed, arranged, ed. and introduction by Dennis Shrubsall (Lampeter: Edwin Mellen Press, 2006).

W. H. Hudson's Letters to R. B. Cunninghame Graham, ed. with an introduction by Richard Curle (Golden Cockerel Press, 1941).

Hudson W. H. and R. Cunninghame Graham, *Two Letters on an Albatross*, ed. with notes by Herbert Faulkner West (Hanover, NH: Westholm Publications, 1955).

Hudson W. H. and P. Sclater, *Argentine Ornithology* (R. H. Porter, 1888–9).

Merryweather, Maud [W. H. Hudson], 'Wanted, a Lullaby', *Cassell's Family Magazine Illustrated* (March 1875), 213–15.

Prof. Enrique M. S. Pedrotti, 'Two Unpublished Letters by William Henry Hudson', *Conceptos*, 77: 1 (January–February / March–April 2002), 27–9.

Primary sources

Payne, John R., *W. H. Hudson. A Bibliography*, foreword by Alfred A. Knopf (Dawson: Archon Books, 1977).

Shrubsall, Dennis, 'Updating Hudson's Bibliography', *English Literature in Transition 1880–1920*, 31: 2 (1988), 186–8.

Wilson, G. F., *A Bibliography of the Writings of W. H. Hudson* (Bookman's Journal, 1922).

Secondary sources

Ackroyd, Peter, *T. S. Eliot* (Hamish Hamilton, 1984).

Aldington, Richard, *Life for Life's Sake: A Book of Reminiscences* (Cassell, 1941).

——, 'The Prose of W. H. Hudson', *Egoist* (15 May 1914), 186–7.

Allen, David Elliston, *The Naturalist in Britain* (Allen Lane, 1976).

Alt, Christina, *Virginia Woolf and the Study of Nature* (Cambridge University Press, 2010).

Ara, Guillermo, 'Guillermo Enrique Hudson: El paisaje pampeano y su experiencia', doctoral thesis (Buenos Aires: Ministerio de Aeronáutica, 1954).

Armaignac, H., *Viajes por las pampas argentinas: Cacerías en el Quequén Grande y otras andanzas, 1869–1874*, tr. from the French *Voyages dans les Pampas*, 1883 (Buenos Aires: Editorial Universitaria de Buenos Aires, 1974).

Arocena, Felipe, *William Henry Hudson: Life, Literature and Science*, trans. Richard Manning (McFarland and Co., 2003).

Assunção, Octavio C., Alicia Haber, Katherine E. Manthorne and Edward J. Sullivan, *The Art of Juan Manuel Blanes* (Buenos Aires: Fundación Bunge y Born, 1994).

Azara, Félix de, *Apuntamientos para la historia natural de los páxaros del Paraguay y Río de la Plata*, 3 vols (Madrid, 1802–5).

——, *Memorias sobre el estado rural del Río de la Plata . . . en 1801* (Madrid, 1847).

B. de C., *A Naturalist and Immortality* (Bellchi Press, 1937).

Backhouse, Hugo, *Among the Gauchos* (Jarrolds, 1950).

Baldwin, Stanley, *On England and Other Addresses* (Harmondsworth: Penguin, 1939).

Barnabé, Jean-Philippe, 'Días de ocio en la Banda Oriental', conference paper (Montevideo, June 2004).

——, *William Henry Hudson y La Tierra Purpúrea: Reflexiones desde Montevideo* (Montevideo: Linardi y Risso, 2005).

Barros, Alvaro, *Fronteras y territorios federales de las Pampas del Sud* (Belgrano: Tipos á Vapor, 1872).

Barroso, Haydée M. Jofre, *Genio y figura de Guillermo Enrique Hudson* (Buenos Aires: Editorial Universitaria de Buenos Aires, 1972).

Bartholomew, Roy (ed.), *Cien poesías rioplatenses, 1800–1950: Antología; Apéndice con los poemas de William Henry Hudson* (Buenos Aires: Editorial Raigal, 1954).

Bassett, Philippa (compiler), *Royal Society for the Protection of Birds, a Digitalised Catalogue by the National Archives: A List of the Historical Records* (RSPB, August 1980).

Bate, Jonathan, *The Song of the Earth* (Picador, 2000).

Bewick, Thomas (illustrator), *A Country Zodiac* (Zodiac Books, 1948).

Birkhead, Tim, *Bird Sense: What it's Like Being a Bird* (Bloomsbury, 2012).

——, *The Wisdom of Birds: An Illustrated History of Ornithology* (Bloomsbury, 2008).

Blunden, Edmund, *Nature in English Literature* (Hogarth Press, 1929).

——,*Undertones of War* (Penguin, 2000).

Blunden, Margaret, *The Countess of Warwick: A Biography* (Cassell, 1967).

Blunt, Wilfrid Scawen, *My Diaries, Being a Personal Narrative of Events, 1888–1914* (Martin Secker, 1920).

Borges, Jorge Luis, 'Sobre The Purple Land', *Obras Completas* (Buenos Aires: Emecé Editores, 1974), 737–9.

Brack, O. M. and James J. Hill, 'Morley Roberts's First Meeting with W. H. Hudson', *English Literature in Transition 1880–1920*, 18: 1 (1975), 36–7.

Brooke, Margaret, *Good Morning & Good Night* (Constable, 1934).

Burton, Capt. Richard, *Letters from the Battle-Fields of Paraguay* (Tinsley Brothers, 1870).

Carman, Raúl L., 'Lugones y las aves', *Nuestras aves* (April 1988), 1–6.

Casares, Jorge, 'Guillermo Enrique Hudson y su amor a los pájaros', *El hornero* (1929), 16–27.

Chatwin, Bruce and Paul Theroux, *Patagonia Revisited* (Salisbury: Michael Russell, 1985).

Church, Richard, *British Authors: A Twentieth Century Gallery* (Longmans Green, 1943).

Cocker, Mark, 'Country Diary', *Guardian* (12 November 2012), 35.

Conrad, Joseph, *Last Essays* (Dent, 1924).

Coustillas, Pierre, *The Heroic Life of George Gissing, Part III: 1897–1903* (Pickering and Chatto, 2012).

—— (ed.), *London and the Life of Literature in Late Victorian England: The Diary of George Gissing, Novelist* (Hassocks: Harvester Press, 1978).

Cunninghame Graham, R. B., *The Conquest of the River Plate* (William Heinemann, 1924).

——, *Hernando de Soto* (William Heinemann, 1912).

——, *The Horses of the Conquest* (William Heinemann, 1930).

——, 'Introduction', W. H. Hudson, *Far Away and Long Ago* (Buenos Aires: Guillermo Kraft, 1943).

——, *Rodeo: A Collection of the Tales and Sketches*, selected by A. F. Tschiffely (William Heinemann, 1936).

——, *The South American Sketches of R. B. Cunninghame Graham*, selected and ed., with an introduction, notes, glossary and bibliography by John Walker (Norman, OK: University of Oklahoma Press, 1978).

——, *A Vanished Arcadia, Being Some Account of the Jesuits in Paraguay, 1607 to 1767* (William Heinemann, 1901).

Curle, Richard, *Caravansary and Conversations: Memories of Places and Persons* (Cape, 1937).

——, 'Introduction', W. H. Hudson, *Birds of La Plata* (Penguin, 1952).

——, 'W. H. Hudson', *Fortnightly Review*, 118 (October 1922), 612–9.

Darwin, Charles, 'Note on the Habits of the Pampas Woodpecker (*Colaptes campestris*)', *Proceedings of the Zoological Society* (November 1870), 705–6.

——, *The Voyage of the Beagle*, facsimile edition (Geneva: Heron Books, no date [1845]).

Darwin, Francis (ed.), *His Life Told in an Autobiographical Chapter, and in a Selected Series of His Published Letters* (John Murray, 1908).

Davies, W. H., *Later Days* (Cape, 1925).

Deguiseppe, Alcides (ed.), *Hudson en Quilmes y Chascomús* (La Plata: Cuadernos del Instituto de Literatura, 1971).

de la Mare, Walter, *Pleasures and Speculations* (Faber, 1940).

Di Giacomo, Alejandro, 'Nidificación del Chilfón (*Syrigma sibilatrix*) en Salto, Buenos Aires, Argentina', *El Hornero*, 13 (November 1988), 1–7.

Dudgeon, Patrick, 'Nature and the Countryside in Argentina and England: W. H. Hudson and Robert Bloomfield', in Nadine Aguilar (ed.) *Aspectos de la obra de W. H. Hudson* (Buenos Aires, 1992), 17–44.

E., C. S., 'Adventures among Books: W.H. Hudson', *Teachers World* (30 November 1921).

Echeverría, Esteban, *Slaughter Yard*, trans. Norman Thomas di Giovanni and Susan Ashe, edited with an introduction and notes by Norman Thomas di Giovanni (Friday Books/HarperCollins, 2010).

Eliot, T. S., 'Rudyard Kipling', *A Choice of Kipling's Verse* (Faber, 1942).

Epstein, Jacob, *An Autobiography* (Hutton Press, 1955).

Fariñas, Analía Hebe, 'Los Hudson en Florencio Varela', *9 Jornada de Estudios sobre el partido de Almirante Brown: Educación y Cultura* (2012), 16–40.

Fernández, Laura, 'La pampa de memoria. William H. Hudson', *Ciber Letras*, 22 (July 2014), http://www.lehman.cuny.edu/ciberletras/v09/fernandez.html.

Ferreyra, Ana Inés, 'La organización de la propriedad en la provincia de Córdoba, siglo xix', *Revista de Investigación* (July 2009), http://www.redalyc.org/pdf/2791/279122165007.pdf.

Finch, Edith, *Wilfrid Scawen Blunt, 1840–1922* (Cape, 1938).

Ford, Ford Madox, *England and the English*, ed. with an introduction by Sara Haslam (Carcanet, 2003).

——, *Letters of Ford Madox Ford*, ed. Richard M. Ludwig (Princeton: Princeton University Press, 1965).

——, 'A Naturalist from La Plata. W. H. Hudson', in *Memories and Impressions*, selected and introduced by Michael Killigrew (Penguin, 1979), 261–70.

——, *Return to Yesterday: Reminiscences 1894–1914* (Victor Gollancz, 1931).

——, 'W. H. Hudson', *English Review* (1913), 3–12.

——, 'W. H. Hudson, Some Reminiscences', *Little Review: A Magazine of the Arts –Making No Compromise with Public Taste*, 7: 1 (May–June 1920), 1–12.

Franco, Jean, 'Prólogo', in Guillermo Enrique Hudson, *La tierra purpúrea/Allá lejos y hace tiempo*, tr. Idea Vilariño (Caracas: Biblioteca Ayacucho, 1980), ix–xlv.

Franco, Luis, *Hudson a caballo* (Buenos Aires: Alpe, 1956).

——, *La pampa habla*, preliminary study by Daniel Campione (Buenos Aires: Ediciones Biblioteca Nacional, 2008).

Franklin, Benjamin, *Autobiography and Other Writings* (Oxford: Oxford University Press, 2008).

Frederick, J. T., *W. H. Hudson* (New York: Twayne, 1972).

Friedman, Terry, *Epstein's Rima. 'The Hyde Park Atrocity; Creation and

Controversy' (Leeds: Henry Moore Centre/Leeds City Art Galleries, 1988).

Fry, Roger, 'The Hudson Memorial', *Dial* (November 1925), 370–73.

Gardiner, Linda, 'Origin of the Hudson Memorial Rima, Hyde Park, London', in Samuel Looker (ed.), *W. H. Hudson: A Tribute by Various Writers* (Worthing: Aldridge Bros, 1947), 147–8.

——, 'W. H. Hudson', *Bird Notes and News*, 3 (Autumn 1922), 33–7.

——, 'W. H. Hudson and the Birds', *Microcosm*, 9: 4 (Winter 1925), 17.

Garnett, David, *Great Friends: Portraits of Seventeen Writers* (Macmillan, 1979).

Garnett, Edward, *Friday Nights: Literary Criticism and Appreciations* (Cape, 1929).

Gatrell, Simon, 'The Letters of W. H. Hudson', *English Literature in Transition 1880–1920*, 51: 3 (1994), 302–14.

Gómez, Leila (ed.), *La piedra del escándalo: Darwin en Argentina (1845–1909)* (Buenos Aires: Ediciones Simurg, 2008).

Gómez, Leila and Sara Castro-Kláren (eds), *Entre Borges y Conrad – estética y territorio en W. H. Hudson* (Madrid: Iberoamericana, 2012).

Gosse, Philip, *Traveller's Rest* (Cassell & Co, 1937).

Gray, John, *The Silence of Animals: On Progress and Other Modern Myths* (Penguin, 2013).

——, *Straw Dogs: Thoughts on Humans and Other Animals* (Granta, 2003).

Grey, Sir Edward, *The Charm of Birds* (Weidenfeld, 2001).

——, *The Cottage Book, the Undiscovered Country Diary of an Edwardian Statesman, Sir Edward Grey*, ed. and introduced by Michael Waterhouse (Weidenfeld & Nicolson, 2001).

——, 'W. H. Hudson, an Appreciation', loose leaf handed out at launch of Jacob Epstein's *Rima* at the W. H. Hudson Memorial, London (1925).

Guthrie, James, 'Edward Thomas' Letters to W. H. Hudson', *London Mercury*, 11: 10 (August 1920).

Hamilton, Robert, *W. H. Hudson: The Vision of Earth* (Dent, 1946).

Harris, P. R., *The British Library, 1753–1973* (British Museum, 1998).

Harvey, Brian, *W. H. Hudson and 'The Great Dog-Superstition'* (Iowa: Shaggy Dog Stories, 2012).

Haymaker, Richard E., *From Pampas to Hedgerows and Downs: A Study of W. H. Hudson* (New York: Bookman, 1958).

Head, Sir Francis B., *Rough Notes Taken During Some Rapid Journeys across the Pampas* (John Murray, 1826).

Hemingway, Ernest, *Hemingway*, ed. Malcolm Cowley (New York: Viking Press, 1944).

Hernández, José, *Martín Fierro*, ed. Luis Sáinz de Medrano (Madrid: Cátedra, 1998).

Herrera, Luis Costa, *Un viaje por la tierra purpúrea* (Montevideo: Ediciones M, 1952).

Hollis, Matthew, *Now All Roads Lead to France: The Last Years of Edward Thomas* (Faber, 2011).

Homenaje a Guillermo E. Hudson (San Miguel de Tucumán: Fundación e Instituto Miguel Lillo, 1972).

Huberman, Ariana, *Gauchos and Foreigners: Glossing Culture and Identity in the Argentine Countryside* (Lexington Books, 2011).

Hunt, Violet, 'The Death of Hudson', *English Review*, 36 (January 1923), 23–5.

——, *The Flurried Years* (Hurst & Blackett, 1926).

Imhoff, Joshua, 'W. H. Hudson: Between Art and Science', MA thesis (Oxford, OH: Miami University, 2009).

Jagoe, Eva-Lyn Alicia, *The End of the World as They Knew it: Writing Experiences of the Argentine South* (Lewisburg, PA: Bucknell University Press, 2008).

James, Henry, *English Hours*, introduction by Leon Edel (Oxford: Oxford University Press, 1981).

Jameson, Storm, *Morley Roberts: The Last Eminent Victorian* (Unicorn Press, 1961).

Jurado, Alicia, *El escocés errante: R. B. Cunninghame Graham* (Buenos Aires: Emecé Editores, 1978).

——, *Vida y obra de W. H. Hudson* (Buenos Aires: Fondo Nacional de las Artes, 1971).

Keen, Jorge, 'W. H. Hudson' (typescript held in the Museo Hudson, Florencio Varela, undated).

Kendall, Ena, 'How a Birdman Feathered his Nest', *Observer Magazine* (23 March 1980), 40–46.

Kernahan, Coulson, *Celebrities: Little Stories about Famous Folk* (Hutchinson, 1923).

Lemon, L. M., 'Linda Gardiner', *Bird Notes & News*, 19 (Spring 1941), 91–3.

Lemon, Mrs Frank, 'W. H. Hudson', *Bird Notes & News*, 19 (Autumn 1941), 129–32.

Liandrat, Francisque, *W. H. Hudson, 1841–1922: Naturaliste. Sa vie et son oeuvre* (Lyon: M. Audin, 1946).

Lombán, Juan Carlos, 'Fernando Pozzo y el descubrimiento de Hudson', *La prensa*, 16 de Julio 1972.

——, *Guillermo Enrique Hudson o el legado inmerecido* (Privately printed, 1986).

——, 'Guillermo Enrique Hudson, un ilustre desconocido', *Nueva Historia de Quilmes*, 128–30 (Quilmes: El Monje Editor, 1992).

Looker, Samuel (ed.), *W. H. Hudson: A Tribute by Various Writers* (Worthing: Aldridge Bros, 1947).

Lugones, Leopoldo, *El payador* (Buenos Aires: Eudeba, 2012).

Lynch, John, 'From Independence to National Organization', in Leslie Bethell (ed.), *Argentina since Independence* (Cambridge: Cambridge University Press, 1993).

Mabey, Richard, *Gilbert White* (Century Hutchinson, 1986).

MacCann, William, *Two Thousand Miles Ride through the Argentine Provinces, with a Historical Retrospect of the Rio de la Plata, Montevideo and Corrientes* (New York: AMS Editions, 1971).

MacDonagh, Emiliano, 'Recuerdos de Hudson en Patagones', *La Nación* (8 May 1932).

——, 'Sir James Mackenzie-Davidson, un argentine descubierto por Hudson', *La Nación* (23 December 1928), 8.

Martínez Estrada, Ezequiel, *El mundo maravilloso de Guillermo Enrique Hudson* (Buenos Aires: FCE, 1951).

Masefield, John, *Rosas* (New York: Macmillan, 1918).

——, *So Long to Learn: Chapters of an Autobiography* (Heinemann, 1952).

Mason, Michael, 'O Cruel!', *London Review of Books* (16 June–6 July 1983), 20–22.

Massingham, H. J., 'W. H. Hudson', *Untrodden Ways* (Fisher Unwin, 1923).

Mehta, Ved, *Mahatma Gandhi and His Apostles* (Penguin, 1977).

Miller, David, *W. H. Hudson and the Elusive Paradise* (Macmillan, 1990).

Moncaut, Carlos Antonio, *Reminiscencias del gaucho Guillermo Enrique Hudson y brevario de sus Pájaros del Plata* (La Plata: privately printed, 1961).

Moreno, Francisco Pascasio, *Viaje a la Patagonia Austral* (Buenos Aires: La Nación, 1879).

Mulhall, M. G. and E. T., *Handbook of the River Plate* (Trübner, 1885).

Musters, George, *At Home with the Patagonians: A Year's Wanderings over the Untrodden Ground from the Straits of Magellan to the Rio Negro* (John Murray, 1871).

Narosky, Tito, *Aves argentinas* (Buenos Aires: Editorial Albatros, 1985).

Neacey, Markus, 'Morley Roberts' Literary Career in the 1880 and 1890s', *Gissing Review*, 68 (April and July 2012).

O'Rell, Max [Paul Blouet], *John Bull and His Island* (London, 1883).

Pedrotti, Enrique M. S., 'Los Hudson en Argentina', *Conceptos*, 71: 4 (July–August, 1996), 22–4.

Peñaloza, Fernanda, Jason Wilson and Claudio Canaparo, *Patagonia Myths and Realities* (Peter Lang, 2010).

Pound, Ezra, 'Hudson: Poet Strayed into Science', *Selected Prose 1909–1965* (Faber, 1973).

Pozzo, Fernando (ed.), *Antología de G. E. Hudson, precedida de estudios críticos sobre su vida y su obra* (Buenos Aires: Losada, 1941).

R. E., 'The Work of W. H. Hudson', *English Review* (April 1909), 157–64.

Renier, G. J., *The English: Are They Human?* (Leipzig: Bernhard Tauchnitz, 1932).

Rich, Paul, 'The Quest for Englishness', *History Today*, 37: 6 (1987).

Roberts, Morley, 'W. H. Hudson', *Cornhill Magazine*, 69 (October 1930), 406–18.

——, *W. H. Hudson: a Portrait* (New York: E. P. Dutton, 1924).

Rodker, John, 'W. H. Hudson', *Little Review: A Magazine of the Arts – Making No Compromise with Public Taste*, 7: 1 (May–June 1920).

Ronner, Amy, *W. H. Hudson, the Man, the Novelist, the Naturalist* (New York: AMS Press, 2006).

Rothenstein, William, *Men and Memories, Recollections of William Rothenstein, 1900–1922* (Faber, 1932).

——, *Twenty-four Portraits* (George Allen, 1920).

Salt, Henry, 'W. H. Hudson, as I Saw Him', in *The Company I Have Kept* (George Allen, 1930), 117–29.

Sasturain, Juan, 'El insoslayable Hudson', *Página 12* (15 October 2012).

Sarramone, Alberto, *Catriel y los indios pampas de Buenos Aires* (Azul: Editorial Biblos, 1993).

Sastre, Marcos, *El Tempe argentino* (Buenos Aires, 1858).

Schmitt, Cannon, *Darwin and the Memory of the Human: Evolution, Savages and South America* (Cambridge: Cambridge University Press, 2009).

Shakespeare, Nicholas, 'Afterword', in W. H. Hudson, *Far Away and Long Ago* (Eland, 1982), 333–7.

Shrubsall, Dennis, *Unpublished Letters of W. H. Hudson . . .*, collected, transcribed, arranged, edited and introduced by Dennis Shrubsall, vols 1–2 (Lampeter: Edwin Mellen Press, 2006).

——, *Walking with W. H. Hudson through the English Landscape: The Home Country of the World's First Literary Environmentalist* (Lampeter: Edwin Mellen Press, 2008).

——, *W. H. Hudson: Writer and Naturalist* (Salisbury: Compton Press, 1978).

Shrubsall, Dennis and Pierre Coustillas, *Landscapes and Literati: Unpublished Letters of W. H. Hudson and George Gissing* (Salisbury: Michael Russell: Salisbury, 1985).

——, *The Writings of W. H. Hudson, the First Literary Environmentalist, 1841–1922: A Critical Survey* (Lampeter: Edwin Mellen Press, 2007).

Shrubsall, Dennis and Martha Vogeler, *The Private Reflections and Opinions of W. H. Hudson* (Lampeter: Edwin Mellen Press, 2008).

Slatter, Richard, *Gauchos and the Vanishing Frontier* (Nebraska: University of Nebraska Press, 1922).

Spencer, Colin, *The Heretic's Feast: A History of Vegetarianism* (Fourth Estate, 1993).

Stoneman Douglas, Marjory, W. H. Hudson papers (Coral Gables, FL: Richter Library, Special Collections, University of Miami Library).

Tesler, Mario, *Camila y la Bemberg: Del Socorro a Pilar. Tragedia y ficción cinematográfica* (Buenos Aires: Biblioteca Nacional, 2010).

Thomas, Edward, *Selected Letters*, ed. R. George Thomas (Oxford: Oxford University Press, 1995).

——, 'W. H. Hudson', in *A Literary Pilgrim*, ed. with an introduction by Michael Justin Davis (Exeter: Webb and Bower, 1985).

Thomas, Helen, *A Memory of W. H. Hudson*, introduction by Myfanwy Thomas, illustrations by Michael Renton (Fleece Press, 1984).

Thomas, Keith, *Man and the Natural World: Changing Attitudes in England 1500–1800* (Penguin, 1984).

Thomson, Hugh, *The Green Road into the Trees: An Exploration of England* (Preface, 2012).

Tippett, Brian, *W. H. Hudson in Hampshire*, Hampshire Papers 27 (Basingstoke: Hampshire County Council, 2004).

Tomalin, Ruth, *W. H. Hudson* (H. F. and G. Witherby, 1954).

——, *W. H. Hudson: A Biography* (Faber, 1982).

Tomlinson, Philip, 'W. H. Hudson (1841–1922): A Modern Worshipper of Nature – Wild Life in Two Worlds', *Times Literary Supplement* (2 August 1941).

Trevelyan, George Macaulay, *Grey of Fallodon being the Life of Sir Edward Grey afterwards Viscount Grey of Fallodon* (Longman Green, 1937).

Tschiffely, A. F., *Don Roberto: Being the Account of the Life and Work of R. B. Cunninghame Graham, 1852–1936* (Heinemann, 1937).

Tsuda, Masao, 'G. E. Hudson y su sobrina Laura', *La Prensa* (14 October 1962).

——, *Las Huellas de Guillermo Enrique Hudson* (Buenos Aires: Américalee, 1963), 13–22.

Velázquez, Luis Horacio, *Guillermo Enrique Hudson: Vida. Obras. Ideas. Magia* (Buenos Aires: Ediciones Culturales Argentinas, 1963).

——, *Hudson vuelve* (La Plata: Ed. Lanura, 1953).

Walker, John, 'From the Argentine Plains to Upper Canada, Sir Francis Bond Head: Gaucho Apologist and Costumbrist of the Pampa', *Canadian Journal of Latin American Studies*, 5: 9 (1980), 97–120.

——, 'Home Thoughts from Abroad: W. H. Hudson's Argentine Fiction', *Canadian Review of Comparative Literature* (September 1983), 333–76.

Wallace, Alfred Russell, *My Life: A Record of Events and Opinions* (London: Chapman & Hall, 1905).

——, Online Correspondence (London: Natural History Museum).

——, 'A Remarkable Book on the Habits of Animals', *Nature*, 45 (14 April 1892), 553–5.

——, 'Reveries of a Naturalist', *Nature*, 47 (23 March 1893), 483–4.

Warwick, Frances Evelyn, *Nature's Quest* (John Murray, 1936).

——, *Life's Ebb & Flow* (Hutchinson, 1929).

Wasson, Gordon and Edwin Way Teal, 'W. H. Hudson's Lost Years', *Times Literary Supplement* (5 April 1947), 161.

Watts, Cedric, *R. B. Cunninghame Graham* (Boston: Twayne, 1983).

West, Herbert Faulkner, *For a Hudson Biographer* (Hanover, NH: Westholm Publications, 1958).

——, *W. H. Hudson's Reading* (Privately printed, 1947).

William Henry Hudson, Commemoration of the 150th Anniversary of his Birth in Quilmes, Argentina (Embassy of the Argentina Republic, 1991).

Williams-Ellis, Annabel, *Darwin's Moon: A Biography of Alfred Russell Wallace* (Blackie, 1966).

Wilson, Jason, 'Charles Darwin and W. H. Hudson', in Alistair Hennessy and John King (eds), *The Land that England Lost: Argentina and Britain, a Special Relationship* (British Academic Press, 1992), 173–82.

——, *W. H. Hudson: The Colonial's Revenge*, Working Papers 5 (Institute of Latin American Studies, 1981).

——, 'W. H. Hudson: Poet in Prose', *Poetry Nation Review*, 32 (1983), 22–6.

Woolf, Virginia, 'Mr Hudson's Childhood', *Times Literary Supplement* (26 September 1918), 453.

Worster, Donald, *Nature's Economy: The Roots of Ecology* (New York: Anchor Books, 1979).

Zorraquín Becú, Horacio, *Tiempo y vida de José Hernández, 1834-1886* (Buenos Aires: Emecé, 1972).

Acknowledgements

I would like to thank Andreas Campomar for his shrewd advice and enthusiasm, and Howard Watson who helped shape the book. I would like to thank Andrea as always for her continuous encouragement. Many thanks for long conversations with Sr Rubén Ravera, director of the Museo Histórico Provincial Guillermo Enrique Hudson and its resident naturalist Marcelo Montenegro; to the librarian Ramiro at the Sociedad Ornitológica del Plata; for long teas with Professor Enrique Pedrotti, director of the Sociedad Amigos de Hudson; to Elizabeth Allen at the Hudson archives at the RSPB library at Sandy, Bedfordshire; Rebecca Bauman for the Curle/Hudson letters at Indiana University Library; Steve Hersh for the Marjory Stoneman Douglas papers at Special Collections, University of Miami Library; to Andrew Gansky for the Knopf/Hudson correspondence at the Harry Ransom Centre, the University of Texas Library at Austin; to Michael Palmer at the Zoological Society Library; and to Lyn Reece at the Central Archives of the British Library for sending me copies of Hudson's signature for his Reader's ticket. Finally, I would like to thank Tom Sutherland, Ricardo Herrera, Miriam Rubino, Alejandro Marara and Professors Pierre Coustillas, Jens Andermann and Jean-Philippe Barnabé.

Image Acknowledgements

The author and publisher wish to express their thanks to the following sources of illustrations and / or permission.

Page 13 Jacob Epstein by his *Rima*, from Terry Friedman, *Epstein's Rima*. *'The Hyde Park Atrocity; Creation and Controversy'*; page 39 W. H. Hudson, from the Smithsonian; page 294 Sir William Rothenstein, from the National Portrait Gallery; page 317 William Lawes, from Ruth Tomalin, *W. H. Hudson: A Biography*; page 290 John Galsworthy, from the Mary Evans Picture Library; page 297 W. H. Davies, page 341 Walter de la Mare, page 275 Ford Madox Ford, from Richard Church, *British Authors. A Twentieth Century Gallery*; page 274 Joseph Conrad, from Time Life Pictures/Getty Images; page 27 1874 Oxen, page 67 Pueyrredón's *Rest in the Country*, page 92 Emeric Essex Vidal's watercolour, page 102 Carlos Pellegrini's *Carro*, page 112 Paueke's *Horsemen Fishing*, page 193 Bacle's *Carnicería*, page 194 Carlos Pellegrini's *El Matadero*, page 195 Bacle's *Corrales de Abasto* from Bonifacio del Carril and Aníbal G. Aguirre Saravia, *Iconografía de Buenos Aires. La ciudad de Garay hasta 1852*; page 82 Juan Manuel Blanes' *La Cautiva*, page 147 Juan Manual Blanes' *La Fiebre Amarilla*, page 168 Juan Manual Blanes' *Los boleadores*, from *The Art of Juan Manuel Blanes*, Fundación Bunge y Born.

The remaining photographs are by Jason Wilson.

Index